AF902496

Clothing for Children and Teenagers

The Textile Institute and Woodhead Publishing

The Textile Institute is a unique organisation in textiles, clothing and footwear. Incorporated in England by a Royal Charter granted in 1925, the Institute has individual and corporate members in over 90 countries. The aim of the Institute is to facilitate learning, recognise achievement, reward excellence and disseminate information within the global textiles, clothing and footwear industries.

Historically, The Textile Institute has published books of interest to its members and the textile industry. To maintain this policy, the Institute has entered into partnership with Woodhead Publishing Limited to ensure that Institute members and the textile industry continue to have access to high calibre titles on textile science and technology.

Most Woodhead titles on textiles are now published in collaboration with The Textile Institute. Through this arrangement, the Institute provides an Editorial Board which advises Woodhead on appropriate titles for future publication and suggests possible editors and authors for these books. Each book published under this arrangement carries the Institute's logo.

Woodhead books published in collaboration with The Textile Institute are offered to Textile Institute members at a substantial discount. These books, together with those published by The Textile Institute that are still in print, are offered on the Elsevier website at: http://store.elsevier.com/. Textile Institute books still in print are also available directly from the Institute's web site at: www.textileinstitutebooks.com.

A list of Woodhead books on textiles science and technology, most of which have been published in collaboration with the Textile Institute, can be found towards the end of the contents pages.

Related titles

Information Systems for the Fashion and Apparel Industry
(ISBN 978-0-08-100571-2)

Garment Manufacturing Technology
(ISBN 978-1-78242-232-7)

Electronic Textiles: Smart Fabrics and Wearable Technology
(ISBN 978-0-08-100201-8)

Woodhead Publishing Series in Textiles:
Number 183

Clothing for Children and Teenagers

Anthropometry, Sizing and Fit

Norsaadah Zakaria

The Textile Institute

ELSEVIER

AMSTERDAM • BOSTON • CAMBRIDGE • HEIDELBERG
LONDON • NEW YORK • OXFORD • PARIS • SAN DIEGO
SAN FRANCISCO • SINGAPORE • SYDNEY • TOKYO
Woodhead Publishing is an imprint of Elsevier

WP
WOODHEAD
PUBLISHING

Published by Woodhead Publishing in association with The Textile Institute
Woodhead Publishing is an imprint of Elsevier
The Officers' Mess Business Centre, Royston Road, Duxford, CB22 4QH, UK
50 Hampshire Street, 5th Floor, Cambridge, MA 02139, USA
The Boulevard, Langford Lane, Kidlington, OX5 1GB, UK

Copyright © 2016 Elsevier Ltd. All rights reserved.

No part of this publication may be reproduced or transmitted in any form or by any means, electronic or mechanical, including photocopying, recording, or any information storage and retrieval system, without permission in writing from the publisher. Details on how to seek permission, further information about the Publisher's permissions policies and our arrangements with organizations such as the Copyright Clearance Center and the Copyright Licensing Agency, can be found at our website: www.elsevier.com/permissions.

This book and the individual contributions contained in it are protected under copyright by the Publisher (other than as may be noted herein).

Notices
Knowledge and best practice in this field are constantly changing. As new research and experience broaden our understanding, changes in research methods, professional practices, or medical treatment may become necessary.

Practitioners and researchers must always rely on their own experience and knowledge in evaluating and using any information, methods, compounds, or experiments described herein. In using such information or methods they should be mindful of their own safety and the safety of others, including parties for whom they have a professional responsibility.

To the fullest extent of the law, neither the Publisher nor the authors, contributors, or editors, assume any liability for any injury and/or damage to persons or property as a matter of products liability, negligence or otherwise, or from any use or operation of any methods, products, instructions, or ideas contained in the material herein.

British Library Cataloguing-in-Publication Data
A catalogue record for this book is available from the British Library

Library of Congress Cataloging-in-Publication Data
A catalog record for this book is available from the Library of Congress

ISBN: 978-0-08-100226-1 (print)
ISBN: 978-0-08-100254-4 (online)

For information on all Woodhead Publishing publications
visit our website at https://www.elsevier.com/

ELSEVIER
Book Aid International
Working together to grow libraries in developing countries
www.elsevier.com • www.bookaid.org

Publisher: Matthew Deans
Acquisition Editor: David Jackson
Editorial Project Manager: Edward Payne
Production Project Manager: Debasish Ghosh
Designer: Victoria Pearson Esser

Contents

Woodhead Publishing Series in Textiles

1 **Watson's textile design and colour Seventh edition**
Edited by Z. Grosicki
2 **Watson's advanced textile design**
Edited by Z. Grosicki
3 **Weaving Second edition**
P. R. Lord and M. H. Mohamed
4 **Handbook of textile fibres Volume 1: Natural fibres**
J. Gordon Cook
5 **Handbook of textile fibres Volume 2: Man-made fibres**
J. Gordon Cook
6 **Recycling textile and plastic waste**
Edited by A. R. Horrocks
7 **New fibers Second edition**
T. Hongu and G. O. Phillips
8 **Atlas of fibre fracture and damage to textiles Second edition**
J. W. S. Hearle, B. Lomas and W. D. Cooke
9 **Ecotextile '98**
Edited by A. R. Horrocks
10 **Physical testing of textiles**
B. P. Saville
11 **Geometric symmetry in patterns and tilings**
C. E. Horne
12 **Handbook of technical textiles**
Edited by A. R. Horrocks and S. C. Anand
13 **Textiles in automotive engineering**
W. Fung and J. M. Hardcastle
14 **Handbook of textile design**
J. Wilson
15 **High-performance fibres**
Edited by J. W. S. Hearle
16 **Knitting technology Third edition**
D. J. Spencer
17 **Medical textiles**
Edited by S. C. Anand
18 **Regenerated cellulose fibres**
Edited by C. Woodings
19 **Silk, mohair, cashmere and other luxury fibres**
Edited by R. R. Franck
20 **Smart fibres, fabrics and clothing**
Edited by X. M. Tao

21 **Yarn texturing technology**
J. W. S. Hearle, L. Hollick and D. K. Wilson
22 **Encyclopedia of textile finishing**
H.-K. Rouette
23 **Coated and laminated textiles**
W. Fung
24 **Fancy yarns**
R. H. Gong and R. M. Wright
25 **Wool: Science and technology**
Edited by W. S. Simpson and G. Crawshaw
26 **Dictionary of textile finishing**
H.-K. Rouette
27 **Environmental impact of textiles**
K. Slater
28 **Handbook of yarn production**
P. R. Lord
29 **Textile processing with enzymes**
Edited by A. Cavaco-Paulo and G. Gübitz
30 **The China and Hong Kong denim industry**
Y. Li, L. Yao and K. W. Yeung
31 **The World Trade Organization and international denim trading**
Y. Li, Y. Shen, L. Yao and E. Newton
32 **Chemical finishing of textiles**
W. D. Schindler and P. J. Hauser
33 **Clothing appearance and fit**
J. Fan, W. Yu and L. Hunter
34 **Handbook of fibre rope technology**
H. A. McKenna, J. W. S. Hearle and N. O'Hear
35 **Structure and mechanics of woven fabrics**
J. Hu
36 **Synthetic fibres: Nylon, polyester, acrylic, polyolefin**
Edited by J. E. McIntyre
37 **Woollen and worsted woven fabric design**
E. G. Gilligan
38 **Analytical electrochemistry in textiles**
P. Westbroek, G. Priniotakis and P. Kiekens
39 **Bast and other plant fibres**
R. R. Franck
40 **Chemical testing of textiles**
Edited by Q. Fan
41 **Design and manufacture of textile composites**
Edited by A. C. Long
42 **Effect of mechanical and physical properties on fabric hand**
Edited by H. M. Behery
43 **New millennium fibers**
T. Hongu, M. Takigami and G. O. Phillips
44 **Textiles for protection**
Edited by R. A. Scott

45 **Textiles in sport**
Edited by R. Shishoo
46 **Wearable electronics and photonics**
Edited by X. M. Tao
47 **Biodegradable and sustainable fibres**
Edited by R. S. Blackburn
48 **Medical textiles and biomaterials for healthcare**
Edited by S. C. Anand, M. Miraftab, S. Rajendran and J. F. Kennedy
49 **Total colour management in textiles**
Edited by J. Xin
50 **Recycling in textiles**
Edited by Y. Wang
51 **Clothing biosensory engineering**
Y. Li and A. S. W. Wong
52 **Biomechanical engineering of textiles and clothing**
Edited by Y. Li and D. X.-Q. Dai
53 **Digital printing of textiles**
Edited by H. Ujiie
54 **Intelligent textiles and clothing**
Edited by H. R. Mattila
55 **Innovation and technology of women's intimate apparel**
W. Yu, J. Fan, S. C. Harlock and S. P. Ng
56 **Thermal and moisture transport in fibrous materials**
Edited by N. Pan and P. Gibson
57 **Geosynthetics in civil engineering**
Edited by R. W. Sarsby
58 **Handbook of nonwovens**
Edited by S. Russell
59 **Cotton: Science and technology**
Edited by S. Gordon and Y.-L. Hsieh
60 **Ecotextiles**
Edited by M. Miraftab and A. R. Horrocks
61 **Composite forming technologies**
Edited by A. C. Long
62 **Plasma technology for textiles**
Edited by R. Shishoo
63 **Smart textiles for medicine and healthcare**
Edited by L. Van Langenhove
64 **Sizing in clothing**
Edited by S. Ashdown
65 **Shape memory polymers and textiles**
J. Hu
66 **Environmental aspects of textile dyeing**
Edited by R. Christie
67 **Nanofibers and nanotechnology in textiles**
Edited by P. Brown and K. Stevens
68 **Physical properties of textile fibres Fourth edition**
W. E. Morton and J. W. S. Hearle

69 **Advances in apparel production**
Edited by C. Fairhurst
70 **Advances in fire retardant materials**
Edited by A. R. Horrocks and D. Price
71 **Polyesters and polyamides**
Edited by B. L. Deopura, R. Alagirusamy, M. Joshi and B. S. Gupta
72 **Advances in wool technology**
Edited by N. A. G. Johnson and I. Russell
73 **Military textiles**
Edited by E. Wilusz
74 **3D fibrous assemblies: Properties, applications and modelling of three-dimensional textile structures**
J. Hu
75 **Medical textiles and biomaterials for healthcare**
Edited by S. C. Anand, J. F. Kennedy, M. Miraftab and S. Rajendran
76 **Fabric testing**
Edited by J. Hu
77 **Biologically inspired textiles**
Edited by A. Abbott and M. Ellison
78 **Friction in textile materials**
Edited by B. S. Gupta
79 **Textile advances in the automotive industry**
Edited by R. Shishoo
80 **Structure and mechanics of textile fibre assemblies**
Edited by P. Schwartz
81 **Engineering textiles: Integrating the design and manufacture of textile products**
Edited by Y. E. El-Mogahzy
82 **Polyolefin fibres: Industrial and medical applications**
Edited by S. C. O. Ugbolue
83 **Smart clothes and wearable technology**
Edited by J. McCann and D. Bryson
84 **Identification of textile fibres**
Edited by M. Houck
85 **Advanced textiles for wound care**
Edited by S. Rajendran
86 **Fatigue failure of textile fibres**
Edited by M. Miraftab
87 **Advances in carpet technology**
Edited by K. Goswami
88 **Handbook of textile fibre structure Volume 1 and Volume 2**
Edited by S. J. Eichhorn, J. W. S. Hearle, M. Jaffe and T. Kikutani
89 **Advances in knitting technology**
Edited by K.-F. Au
90 **Smart textile coatings and laminates**
Edited by W. C. Smith
91 **Handbook of tensile properties of textile and technical fibres**
Edited by A. R. Bunsell
92 **Interior textiles: Design and developments**
Edited by T. Rowe

93 **Textiles for cold weather apparel**
Edited by J. T. Williams
94 **Modelling and predicting textile behaviour**
Edited by X. Chen
95 **Textiles, polymers and composites for buildings**
Edited by G. Pohl
96 **Engineering apparel fabrics and garments**
J. Fan and L. Hunter
97 **Surface modification of textiles**
Edited by Q. Wei
98 **Sustainable textiles**
Edited by R. S. Blackburn
99 **Advances in yarn spinning technology**
Edited by C. A. Lawrence
100 **Handbook of medical textiles**
Edited by V. T. Bartels
101 **Technical textile yarns**
Edited by R. Alagirusamy and A. Das
102 **Applications of nonwovens in technical textiles**
Edited by R. A. Chapman
103 **Colour measurement: Principles, advances and industrial applications**
Edited by M. L. Gulrajani
104 **Fibrous and composite materials for civil engineering applications**
Edited by R. Fangueiro
105 **New product development in textiles: Innovation and production**
Edited by L. Horne
106 **Improving comfort in clothing**
Edited by G. Song
107 **Advances in textile biotechnology**
Edited by V. A. Nierstrasz and A. Cavaco-Paulo
108 **Textiles for hygiene and infection control**
Edited by B. McCarthy
109 **Nanofunctional textiles**
Edited by Y. Li
110 **Joining textiles: Principles and applications**
Edited by I. Jones and G. Stylios
111 **Soft computing in textile engineering**
Edited by A. Majumdar
112 **Textile design**
Edited by A. Briggs-Goode and K. Townsend
113 **Biotextiles as medical implants**
Edited by M. W. King, B. S. Gupta and R. Guidoin
114 **Textile thermal bioengineering**
Edited by Y. Li
115 **Woven textile structure**
B. K. Behera and P. K. Hari
116 **Handbook of textile and industrial dyeing Volume 1: Principles, processes and types of dyes**
Edited by M. Clark

117 **Handbook of textile and industrial dyeing Volume 2: Applications of dyes**
Edited by M. Clark
118 **Handbook of natural fibres Volume 1: Types, properties and factors affecting breeding and cultivation**
Edited by R. Kozłowski
119 **Handbook of natural fibres Volume 2: Processing and applications**
Edited by R. Kozłowski
120 **Functional textiles for improved performance, protection and health**
Edited by N. Pan and G. Sun
121 **Computer technology for textiles and apparel**
Edited by J. Hu
122 **Advances in military textiles and personal equipment**
Edited by E. Sparks
123 **Specialist yarn and fabric structures**
Edited by R. H. Gong
124 **Handbook of sustainable textile production**
M. I. Tobler-Rohr
125 **Woven textiles: Principles, developments and applications**
Edited by K. Gandhi
126 **Textiles and fashion: Materials design and technology**
Edited by R. Sinclair
127 **Industrial cutting of textile materials**
I. Viļumsone-Nemes
128 **Colour design: Theories and applications**
Edited by J. Best
129 **False twist textured yarns**
C. Atkinson
130 **Modelling, simulation and control of the dyeing process**
R. Shamey and X. Zhao
131 **Process control in textile manufacturing**
Edited by A. Majumdar, A. Das, R. Alagirusamy and V. K. Kothari
132 **Understanding and improving the durability of textiles**
Edited by P. A. Annis
133 **Smart textiles for protection**
Edited by R. A. Chapman
134 **Functional nanofibers and applications**
Edited by Q. Wei
135 **The global textile and clothing industry: Technological advances and future challenges**
Edited by R. Shishoo
136 **Simulation in textile technology: Theory and applications**
Edited by D. Veit
137 **Pattern cutting for clothing using CAD: How to use Lectra Modaris pattern cutting software**
M. Stott
138 **Advances in the dyeing and finishing of technical textiles**
M. L. Gulrajani
139 **Multidisciplinary know-how for smart textiles developers**
Edited by T. Kirstein

140 **Handbook of fire resistant textiles**
Edited by F. Selcen Kilinc
141 **Handbook of footwear design and manufacture**
Edited by A. Luximon
142 **Textile-led design for the active ageing population**
Edited by J. McCann and D. Bryson
143 **Optimizing decision making in the apparel supply chain using artificial intelligence (AI): From production to retail**
Edited by W. K. Wong, Z. X. Guo and S. Y. S. Leung
144 **Mechanisms of flat weaving technology**
V. V. Choogin, P. Bandara and E. V. Chepelyuk
145 **Innovative jacquard textile design using digital technologies**
F. Ng and J. Zhou
146 **Advances in shape memory polymers**
J. Hu
147 **Design of clothing manufacturing processes: A systematic approach to planning, scheduling and control**
J. Gersak
148 **Anthropometry, apparel sizing and design**
D. Gupta and N. Zakaria
149 **Silk: Processing, properties and applications**
Edited by K. Murugesh Babu
150 **Advances in filament yarn spinning of textiles and polymers**
Edited by D. Zhang
151 **Designing apparel for consumers: The impact of body shape and size**
Edited by M.-E. Faust and S. Carrier
152 **Fashion supply chain management using radio frequency identification (RFID) technologies**
Edited by W. K. Wong and Z. X. Guo
153 **High performance textiles and their applications**
Edited by C. A. Lawrence
154 **Protective clothing: Managing thermal stress**
Edited by F. Wang and C. Gao
155 **Composite nonwoven materials**
Edited by D. Das and B. Pourdeyhimi
156 **Functional finishes for textiles: Improving comfort, performance and protection**
Edited by R. Paul
157 **Assessing the environmental impact of textiles and the clothing supply chain**
S. S. Muthu
158 **Braiding technology for textiles**
Y. Kyosev
159 **Principles of colour appearance and measurement**
Volume 1: Object appearance, colour perception and instrumental measurement
A. K. R. Choudhury
160 **Principles of colour appearance and measurement**
Volume 2: Visual measurement of colour, colour comparison and management
A. K. R. Choudhury
161 **Ink jet textile printing**
C. Cie

162 **Textiles for sportswear**
Edited by R. Shishoo
163 **Advances in silk science and technology**
Edited by A. Basu
164 **Denim: Manufacture, finishing and applications**
Edited by R. Paul
165 **Fabric structures in architecture**
Edited by J. Ignasi de Llorens
166 **Electronic textiles: Smart fabrics and wearable technology**
Edited by T. Dias
167 **Advances in 3D textiles**
Edited by X. Chen
168 **Garment manufacturing technology**
Edited by R. Nayak and R. Padhye
169 **Handbook of technical textiles Second edition Volume 1: Technical textile processes**
Edited by A. R. Horrocks and S. C. Anand
170 **Handbook of technical textiles Second edition Volume 2: Technical applications**
Edited by A. R. Horrocks and S. C. Anand
171 **Sustainable apparel**
Edited by R. S. Blackburn
172 **Handbook of life cycle assessment (LCA) of textiles and clothing**
Edited by S. S. Muthu
173 **Advances in smart medical textiles: Treatments and health monitoring**
Edited by L. van Langenhove
174 **Medical textile materials**
Y. Qin
175 **Geotextiles**
Edited by R. M. Koerner
176 **Active coatings for smart textiles**
Edited by J. Hu
177 **Advances in braiding technology: Specialized techniques and applications**
Edited by Y. Kyosev
178 **Smart textiles and their applications**
Edited by V. Koncar
179 **Information systems for the fashion and apparel industry**
Edited by T. M. J. Choi
180 **Antimicrobial textiles**
G. Y. Sun
181 **Advances in technical nonwovens**
G. Kellie
182 **Advances in women's intimate apparel technology**
W. Yu
183 **Clothing for children and teenagers: Anthropometry, sizing and fit**
N. Zakaria

Introduction: clothing for children and teenagers

1

1.1 Historical: manufacturing clothing for children and teenagers

Clothing was styled and designed only for adults throughout the 18th century [1]. There were no clothes styled specifically for children during this time because children were considered as mini adults who wore the adult styles [2]. Furthermore, the children's clothing construction was much stronger than the adult clothing so that the clothes could last longer. The clothes were normally made bigger so they could be worn by multiple siblings.

Slowly, starting from the year 1800s until during the sewing machine Industrial Revolution in 1845, there was a change in how clothes were produced. During this time, the sewing machine was introduced and it became possible to mass produce clothes. With this new concept of production, design patterns were introduced for children's clothing [3]. From this point onward, there were different clothing classifications for adults and children. Clothing design for adults was according to the needs and wants of the adults and design for children was applicable only for children. Since style and fashion were just being introduced to children's clothing, manufacturers mostly designed clothing for children as if they had no taste and desires [4]. Generally, parents bought clothes for their children according to what they perceived was good and not because this was clothing that their children wanted [5].

The Industrial Revolution also initiated the development of factory-made clothes instead of custom-made or made-at-home styles [6]. Mass production began when women started going to work and they no longer had the time to sew their own children's clothes. Factory production also made children's clothes much sturdier and more reliable by introducing different types of fasteners like zippers, buttons, and snaps to secure the clothing on the children. In addition, these new fasteners allowed children to become much more independent in being able to dress themselves [7].

Transformations were seen slowly in the beginning of children's clothing manufacturing. But in time, changes began to occur rapidly because of better manufacturing technology. Children's wear saw the transition from handmade clothes sewn at home to being produced in factories as ready-to-wear (RTW) clothing. Those factories made clothing that would be sold to anybody through catalogs; today clothing is sold online where customers can buy items using a computer or they can still order over the telephone.

In the next section, the discussion focusses on the different trends in clothing manufacturing for today's growing children's and teenagers' clothing market.

Clothing for Children and Teenagers. http://dx.doi.org/10.1016/B978-0-08-100226-1.00001-1
Copyright © 2016 Elsevier Ltd. All rights reserved.

1.2 Introduction: manufacturing clothing for children and teenagers

Today the children's clothing business has evolved into one of the most successful clothing manufacturing industries. Clothing manufacturers have been challenged to meet the demands of children and young teenagers for new trends and designs. Manufacturers are geared toward supplying attractive-looking clothing to the retail market to entice these new-age consumers. The setup of the children's clothing market today is fast-moving and stylish. The retail market sells clothing that is inventive and eye-catching, catering to children's senses and desire to have beautiful clothes that are aesthetically pleasing [8].

Children's clothing manufacturers must always stay attuned to the needs of the market. There's been a huge boom in the children's clothing industry, and to be able to compete actively, these are some of the current manufacturing interests.

1.2.1 Fast fashion trending

The fashion manufacturing cycle has changed and become more rapid in order to appeal to consumers for consistent purchases [9]. Formerly the fashion product life cycle ran about 4–6 months. But now, the cycle is only within 4–6 weeks—this trend is known as fast fashion [10]. According to Fletcher, "Fast fashion—low-cost clothing collections based on current, high-cost luxury fashion trends—is, by its very nature, a fast-response system that encourages disposability" [11]. Furthermore, fast fashion has arisen along with the globalization era, technology adoption, the age of consumerism, and also the green era. According to a report in *Forbes*, Zara is the leader in fast fashion retailers, providing the opportunity for young buyers to have continuous renewed looks [12].

Fast fashion enables more buying power among young women who constantly thirst for something new by giving them affordable apparel [13]. According to Claudio, fast fashion was able to create the new "must-haves" to youngsters because of the quick and rapid process of the new developments that assist in the process of ideation, creation, and construction [14]. The fashion products provide the marketplace with different styles aimed mostly at young women. Teens, especially girls, are known to be very excited about having new things to purchase, and since fast fashion is cheap and new, they can buy more [15,16].

1.2.2 Size and fit

For most apparel consumers size and fit are the main criteria for purchasing ready-to-wear clothes [17,18]. Especially, the clothing worn by children is expected to fit the shapes and sizes of individuals so that the wearer feels comfortable. The comfort level is anticipated to be high for children with different levels of growth [19,20]. The same-aged children can vary in growth rate depending on different factors like diet, genetics, social and environmental factors, etc. The variation in growth patterns will result in variations of body sizes and shapes [21]. Without having a good sizing system that caters to all of the different size ranges it will eventually be difficult to choose the right clothing that fits [22,23]. This topic will be discussed in depth in Chapter 2 as the understanding of body sizes and shapes is crucially needed to develop a good sizing system. Understanding of the sizes and shapes is made possible by conducting an anthropometric survey [24].

1.2.3 Teenagers: a new clothing size group

Teenagers today are very unique. They want to be classified as their own group of consumers [25]. Therefore, the need to customize a different size chart totally on teen sizes is demanded. Previously, only three different size groups were considered: babies, children, and adults. Currently, teenagers no longer want to be clustered under children but instead want to be seen buying in their own section of a store, which is an additional size to the clothing grouping [21]. Teenagers feel that they are becoming more adventurous at selecting their clothes and care a great deal about their appearance and comfort like adult consumers do [26]. By having this new size label specifically for teens, this will stress more-sophisticated expression than buying under a label for a girl's or boy's size [27]. However, the tweens are still sized under children groups. Studies also have shown that there is an increased number of obese young people who require another size chart [28]. There are still scarce numbers of plus size retail shops for tweens and teens carrying attractive assortments of clothing. Some retailers do offer plus size clothing but with limited and unattractive selections [29].

1.2.4 Conscious consumers

Today the demand for clothing manufacturing is really high among the contract manufacturing countries. According to one report, the fashion industry is one of the most important sectors in the world that creates jobs and clothes for people all over the world [30]. Although this sector has created work for millions of people, much of the work is still done under unacceptable conditions and the wages are extremely low [31].

The competition for manufacturing is great among the countries with large populations like China, Indonesia, and India and also the underdeveloped countries such as Bangladesh, Cambodia, Vietnam, Ghana, etc. The demand for clothing to be produced in these countries is due to cheap labor. With all the news about these issues globally, people today are more particular about choosing their clothes, wanting to know where they are made and in what conditions [32]. Furthermore, the tech-savvy children of today are able to understand all these issues and able to get the information from all sources of media. They are very exposed to the latest news such as bad working conditions, child labor manufacturing, low wages, and also slum working areas. Hence, consumers are moving toward more innovative eco-fashion manufacturers [33].

1.2.5 Unique design features

The clothing for children and teenagers not only needs to be attractive but also have a primary focus on comfort and safety [20,34]. Moreover, functional design has started to appear in the children's clothing market [35]. Some of the top functional elements are easy care, extra-long hems and cuffs for hand-me-down or second-life concepts. Attention should be given to the seam selection for certain active parts of the garments for better quality garments. Tweens and teens are perceived as the dynamic clothing group. Thus, their clothing should be made practical, comfortable and use high-tech materials like highly absorbent, ecologically friendly fabrics, and chemically safe dyes. The quality aspects are mandatory elements in children's garments as the competition is rising high. Quality is no longer expected only in high-priced garments but also in low-cost garments due to today's very demanding consumers [36].

1.3 The social psychology of clothing for children and teenagers

In this section, social and psychology aspects are discussed to understand how children and teenagers use their clothing's appearance to express how they perceive of themselves. The social and psychology aspects that will be explored in this chapter are pertinent to the context of body and clothing since the main topic of this book is about understanding body sizes and shapes for clothing sizes. In broader terms, we are focusing on how clothing and the human body will affect the appearance of an individual. Clothing appearance is the way clothing interacts with people, and how it influences them within the society they live in which will give them the sense of self concept or/and self-image [37].

The two different groups of children we are focusing on in this book are the preadolescent and adolescent. Preadolescents are known as "tweens" and adolescents are known as "teens." These two groups are known to be highly affected by psychology and social environment [38].

Tweens are children aged 8–12. They are preadolescent children who want to be teens but still act like kids; knowing how to get their attention is a critical first step [39]. The next group, teens, are children aged 13–19. From the perspective of body growth, both groups are experiencing rapid hormonal changes transitioning from elementary to middle school and then reaching high school [40]. During the period of becoming tweens, they are approaching puberty [41]. With the onset of puberty, this is a period where they will experience great physical, psychological, and social growth [42] as their bodies starts to develop into young adults. The immense changes are a time of transition from childhood to adulthood that consists of many physical, cognitive, and socioemotional changes [43].

In this chapter, relating the rapid physical changes of the groups to the sociopsychology of clothing will lead to the reasons why clothes are worn. But before we explore the reasons why we wear clothing, let's define the word *clothing*, which refers to any tangible or material objects connected to the human body. Clothing are those objects attached to or worn on our bodies and obtained by buying, receiving, or constructing [37]. In addition, appearance is the way a person looks to other people created by a human body or coverings of the body.

The early 19th century marks the start of the Industrial Revolution. The new intellectual and social movement that began during this era caused an active discussion of social and ethical issues related to clothing [44]. The feminist movement was the one that fought to recondition women's clothing by promoting styles that were more functional, less restrictive, and more comfortable than the styles of that era [45]. During this era, children's clothing was also constructed to become simpler and with comfortable styles as compared to the complicated styles in the earlier centuries. The reformation of children's clothing signaled that clothing was becoming more practical and functional for increasingly active school-aged children [46].

It is also noted from the history of clothing that children are always struggling to be separated from their parents in the sense of clothing appearance. After the 19th century, this group of children strove to express their independence through clothing appearance when their clothing styles began to differ significantly from their parents.

In 1952 James Laver [47] wrote in his book *Clothes* of three reasons why clothes were worn:

1. Utility—to protect from weather harshness and for modesty;
2. Hierarchical principle—to lend social status;
3. Seduction principle—to make the wearer as attractive as possible within the framework of what is considered attractive at any particular time in history.

Looking at the reasons for wearing clothing, hereafter it can be connected to the sociopsychology perspectives.

1.3.1 Clothes express peoples' attitudes toward themselves and their society—both what they wear and how they wear it

Wearing clothes that reflects one's identity is acknowledged as the most important expression within the children [48]. The clothes not only reflect who they are but also help them to gain acceptance within their society. It has been found from the research that clothing that does not fit the group's standards may be criticized by means of the person being talked about, laughed at, or teased [49]. During the tween years, children start to look for their identity. They start to discover their own self-esteem that they will use to pass into their teen and young adult years. With positive self-esteem, they will build a good self-concept and self-image [50]. All three factors will create an impact toward developing a good basis of self in the society they are living in. With the sharing of common interests and values, children will grow to relate to clothing importance and fashion interests in their society. At this point of life, clothing starts to play a role in the socialization process [51].

For tweens, parents are seen as the biggest influence on what they are wearing and how it is supposed to be worn [52]. Young tweens look up to their parents for opinions on fashion choices. As the tweens grow older, parental influence on clothes starts to decline [52]. In addition, for the teens, the society also plays a bigger role in influencing what they wear and how clothing makes them feel. Nonetheless, the society of the teens is not the parents anymore but the peers, which are highly regarded as their reference groups. These groups are recognized to have great impact on their decisions [53]. The concept of consumer socialization was acknowledged as the key role in peer group influence [54].

1.3.2 Clothes can be worn to try to change an existing condition

Previous research has confirmed that fashionable clothing is defined as a form of "collective behavior and as a means of social conformity" [55]. Clothes can be worn to influence others and get the others to engage themselves so that they will be identified with their peers. They can also change their clothing preferences according to different conditions in the society [17,56]. As mentioned earlier, clothing was used to reflect oneself and also to gain acceptance by society. As the child becomes confident with high self-esteem, clothes in turn can be used as an appearance to change the perception of others about them [57].

Teens are acknowledged to be very prone to peer pressure but it has been found that boys are less susceptible to peer pressure compared to girls [58]. Young girls are sensitive to group demands and also have a need to be noticed in wearing more brand label clothing [59]. Boys can be seen focusing only on certain brand labels and names that bring significant cues among peers. In addition, girls are also very conscious of the latest trends [18]. They look forward to what's current and changes in trends will make them very trendy, thus encouraging them to buy more clothing to suit that look [60].

Conformity to one's surroundings is very important within these young groups of tweens and teens. By conforming to the society, certain elements can be avoided such as ridicule and social isolation. Besides, conforming to both positive peer pressure and one's school community can bolster a teen's self-confidence [20,61].

1.3.3 Clothes are selected to fulfill the desire to be attractive and fit the parts of the body for a certain appearance

According to one study, young children desire to have a good appearance when wearing clothing. In general, this appearance also includes physical appearance to be attractive to their own societies. They are affected by the social influence and self-perception relative to their group [62,63]. In one study that was done, it was found that when children go through a positive environment and society, their self-esteem increased. Thus this will intensify their awareness about their physical looks in relation to clothing and appearance [64]. The perception about the physical self is known as the body image.

When one views how satisfied they are about their body image, this is called the body cathexis. In one study, it has been discovered that apparel can be used to enhance the feelings of the wearer's ideal physical self [65]. Therefore, providing comfortable and well-fitting clothing to children is imperative. Adult consumers' satisfaction with fit of clothing is linked to store patronage, clothing attitudes, and fashion interests and may provide some insight into the children's market [66,67].

Some research reveals that children select criteria like fit and comfort when selecting what clothes to buy and wear [68,69]. Fit here is related to the conformity of clothes to the body. The children must be satisfied with their body first before choosing the right fit of clothing. Once they are satisfied with their body image, they will choose the right-sized clothing for their body size. Thus choosing the right-fit clothing is vital for this group of children. Regarding teenagers, clothing has been identified as one of the most important contributors for creating better body shape/appearance [39]. They tend to utilize clothes to change their body's perceived shape, which is defined by culture. Conformity of reference group validates the teenagers' opinion and behaviors [70]. Teenagers also assigned similar degrees of importance to several clothing characteristics including aesthetics/design, comfort, fit, and use of socially responsible construction/production method [71].

1.4 The rise in consumption of clothing for children and teenagers

As Waksler [72] noted, in everyday life adults take for granted that children as a category know less than adults, have less experience, are less serious, and are less important than adults in the ongoing work of life. In today's world, Marten [73] added that long gone are the days of indifference to children's voices. They are no longer the passive recipients of adults' needs and culture but are competent social actors—social beings in their own right.

The youngest generation today is known as Generation Z (iGen, Post millennials, Centennials), who were born after the millennials (the Post-Millennials). These are the generation with birth years ranging from 2005 to present time. According to Peterson, they have been identified as traditional retail's worst nightmare [74]. This generation has traits such as entrepreneurial minds, are hyper-aware, multi-tasking, and tech reliant making it a challenge for marketers to understand how to penetrate them [75]. Although they are still young to have incomes, they have significant control over what they purchase such as toys, groceries, and clothing. According to a current study, they have an average allowance of nearly $70 a month, which translates to $44 billion a year for spending alone [76].

These days, the children's clothing market is reported to be one of the most lucrative segments in the global apparel industry. Industry research undertaken by Mintel, UK [77] recognizes children as a key consumer sector, worthy of its own annual report for children's wear, which showed profit growth by 7.7% in the last two years to reach £5.6 billion. Supported by a growing population of children, the children's clothing category in the United States had estimated sales of USD 30.6 billion since 2007 [78]. This market segment is expected to grow 12% in current terms from 2008 onward. According to Cotton Incorporated's Lifestyle Monitor™ (USA), 56% of teenagers ages 13–15 select all or most of their own clothing, as do 30% of those children ages 10–12, and 15% of those children ages 6–9 [79]. PLC, a leading British children's clothing retailer, gave a total of £4.3 billion spent in 2002 alone [80]. Business Wire in France projected that children's wear retail sales within the top 10 global retail markets reached a profit of USD 131.5 billion in 2012 [80]. In 2017 the children's wear market is expected to hit a value of USD 173.6 billion [81].

The children's and teenagers' market has become increasingly significant as children are reported to exert a considerable influence in purchase decisions such as clothing [82,83]. Studies by Cook [84] showed that children are now recognized as "real customers," a privilege formerly overshadowed by their parents. In another study, it was acknowledged that children know what they want and therefore manufacturers and marketers are designing clothes with children's needs and desires in mind [85]. This is called "pediocularity," which means from the child's point of view. Jamaluddin [86] agreed with this statement when his research revealed that parents make no contribution to the decision-making orientation of adolescents.

In addition, a few studies in the literature show that teenage girls buy clothing more often than teenage boys, spending 72% of their income on clothing and clothing-related accessories compared to 52% for boys [87]. On the other hand, tweens are seen as fickle customers and complicated. They are no longer interested in playing with toys as compared to the younger children 5–9 years old. However, they are motivated to spend their own money, which amounted to almost USD 200 billion in sales last year [75]. With the bigger prospect of spending on their disposable income, these groups of youngsters who are worth USD 20 billion are acknowledge to be the new power players of the consumerism. Thus marketers are planning their marketing concept to earn the benefits of an estimated USD 17 billion a year [88]. It is also worth noting here that teens are the most powerful force in the US market now, which spent USD 112.5 billion in 2003 alone [75]. Children's requirements and wishes are regarded as authentic and are now worth a large portion of the clothing industry's attention because they are spending more on clothing than at any other time in history [87]. This is supported by the Piper Jaffray semiannual business report released in mid-2015, from a survey of 6200 teens stating that 20% of teens' income is now spent on fashion clothing [89].

1.5 Conclusions and future trends

The market has definitely recognized children as an important consumer group when it comes to fashion and clothing. This means that the entire apparel market must change and adjust toward the constant and changing values of this young population: the tweens and teens generation. According to Seides in his article "Understanding the teen market and how to maximize your swag," two keys to success are speed to market and responsiveness [83]. Manufacturers and retailers need to withstand these fickle markets. The fast fashion retailers need to figure out where trends are going tomorrow as the teen market has changed significantly.

An understanding of teenagers' buying behavior, and specifically of what teenagers actually value when apparel purchase decisions are made, are of great importance to marketers. These young consumers belong to Generation Z that "has high self-esteem, most notably a savvy with technology and a comfort-level with the global world" [90]. The tween years are deemed a time whereby peer pressure and "fitting in" play a very important part of their childhood. On the other hand, adolescent children are highly affected by the latest fashion trends. Thus the role of the manufacturers is seen to be crucial in order to understand the behavior and attitudes of the young shoppers so that their needs can be fulfilled [91,92].

Finally, it could be concluded that even though internal and external factors do play a role in the selection of clothing, the young consumers agreed that physical body image does influence their appearance when wearing clothes. Body cathexis mediates teenagers' behavior toward viewing their selection of clothes. It is viewed that when they purchase clothing, size and fit of the clothes is regarded as a factor for comfort [39,71]. In order to wear comfortable clothes, sizes based on the body must be collected and understood by the manufacturers. With the right sizing system, the production of the clothing can be derived by these factors.

Future research should explore field research and gather observations related to the paradox of tween and teens' dressing—the exploration of what fashions they like versus the importance of dressing similarly to members of their peer groups. The power of those influencing them, such as parents and peers, over fashion choices should be further investigated. Are the children aware of the fashion gatekeeper role that the environment plays? We should also explore factors that trigger discussions regarding clothing that make fashion statements as well as fashion that contributes to the functional aspects of size and fit.

References

[1] C.R. Callahan, History of Children's Clothing. Available from: http://fashion-history.lovetoknow.com/fashion-history-eras/history-childrens-clothing (retrieved 03.12.15).

[2] L. Baumgarten, Children's Clothing. Available from: http://www.history.org/history/clothing/children/child01.cfm (retrieved 31.12.15).

[3] A. Buck, Clothes and the Child: A Handbook of Children's Dress in England, 1500–1900, Holmes and Meier, New York, 1996.

[4] R. Clare, Children's Clothes Since 1750, Drama Book Publishers, New York, 1989.

[5] M.K. Brock, P.V. Ulrich, L.J. Connell, Exploring the apparel needs and preferences of tween girls' and their mothers, Cloth. Text. Res. J. 28 (2) (2010) 95–111.

[6] S. Benson, J.Y. Stock, Development of the Industrial. U.S. Almanac, UXL, New York, 2005.

[7] How Children's Clothing is Made-Material and Production. Available from: http://www.madehow.com/volume-4/children-s-Clothing.html (retrieved 25.12.15).

[8] J.U. Mc Neal, The Kids Market: Myths and Realities, Paramount Market Publishing, 1999.

[9] A. Joy, J.F. Sherry Jr., A. Venkatesh, J. Wang, R. Chan, Fast fashion, sustainability, and the ethical appeal of luxury brands, Fashion Theory 16 (3) (2012) 273–296.

[10] K. Fletcher, Fashion Needs and Consumption. In Book Titled: Sustainable Fashion & Textiles: Design Journeys, Earthscan, Oxford, 2014, pp. 137–150.

[11] P. Rosenblum, Wet Seal Whipped: Expect More Casualties as Fast Fashion Retailers Take Over Teen Market, January 20, 2015. Available from: http://www.forbes.com/sites/paularosenblum/2015/01/20/fast-fashion-retailers-are-taking-over-the-teen-market/ (retrieved 20.12.15).

[12] W. Loeb, Zara leads in fast fashion. Forbes. Available from: http://www.forbes.com/sites/walterloeb/2015/03/30/zara-leads-in-fast-fashion/March 30, 2015 (retrieved 30.12.15).

[13] FT.com, Fashion: A Better Business Model, June 18, 2014.

[14] L. Claudio, Waste couture: environmental impact of the clothing industry, Environ. Health Perspect. 115 (9) (2007) A449–A454.

[15] S.K. Kaushal, A survey of teenagers' apparel purchase behavior: fashion apparels, promotion, reference group and body cathexis, Paradigm 15 (1) (January 2011) 45–64.

[16] G. Smith, Tweens R' Shoppers: A Look at the Tween Market & Shopping Behavior, March 2, 2013. Available from: http://www.popai.com/uploads/downloads/POPAIWhite-Paper-Tweens-R-Shoppers-2013.pdf (retrieved 06.01.16).

[17] C. Bouchez, Finding clothes that fit and flatter, WebMD Weight Loss Clinic. Available from: http://www.webmd.com/skin-beauty/guide/findingclothes-that-fit-flatter (retrieved 12.06.12).

[18] H.M. De Klerk, T. Tselepis, The early-adolescent female clothing consumer: expectations, evaluation and satisfaction with fit as part of the appreciation of clothing quality, J. Fashion Marketing Manag. 11 (3) (2007) 413–428.

[19] S. Gill, Improving garment fit and function through ease quantification, J. Fashion Marketing Manag. 15 (2) (2011) 228–241.

[20] C.M. Daters, Importance of clothing and self-esteem among adolescents, Cloth. Text. Res. J. 8 (3) (1990) 45–50.

[21] L.J. Connell, P.V. Ulrich, E.L. Brannon, M. Alexander, A.B. Presley, Body shape assessment scale: instrument development for analyzing female figures, Cloth. Text. Res. J. 24 (2) (2006) 80–95.

[22] J.E. Croney, An anthropometric study of young women fashion students including a factor analysis of body measurements, Man 12 (3/4) (1977) 484–496.

[23] E. Bye, K.L. LaBat, M.R. Delong, Analysis of body measurement systems for apparel, Cloth. Text. Res. J. 24 (2) (2006) 66–79.

[24] T. Tselepis, H.M. de Klerk, Early adolescent girls' expectations about the fit of clothes: a conceptual framework, J. Family Ecol. Consum. Sci. 32 (1) (2004) 83–92.

[25] C. Ebster, U. Wagner, D. Neumueller, Children's influence on in-store purchases, J. Retail. Consum. Serv. 16 (2009) 145–154.

[26] D. Prince, N. Martin, The tween consumer marketing model: significant variables and recommended research hypotheses, Acad. Marketing Stud. J. (July 2012).

[27] S. Linn, Consuming Kids, The New Press, New York, NY, 2004.

[28] L.J. Connell, P. Ulrich, Apparel Product Development for the Plus-Size Tween and Teen Market (Project No. S04-AC01), 2006. Available from: http://www.human.cornell.edu/che/fsad/research/ntc.cfm (retrieved 12.06.12).

[29] R.E. Cornette, The emotional impact of obesity on children, in: D. Bagchi (Ed.), Global Perspectives on Childhood Obesity Current Status, Consequences and Prevention, Academic Press. Cotton Inc., London, October 2011, pp. 257–264.

[30] F. Rudell, Shopping with a social conscience: consumer attitudes toward sweatshop labor, Cloth. Text. Res. J. 24 (2006) 282–296.

[31] K. Reinhard, D. Schmidt, F. Rützel, M. Zentgraf, Working conditions in the global fashion industry. Available from: https://laboureconomics.wordpress.com/2013/04/30/working-conditions-in-the-global-fashion-industry/, posted on April 2013 (retrieved 31.12.15).

[32] E. Lewis. In the global apparel industry, abusive and deadly working conditions are still the norm. Available from: http://inthesetimes.com/working/entry/18066/out_of_sight_erik_loomis, posted on June 15, 2015 (retrieved 20.11.15).

[33] T.F. Mangleburg, P.M. Doney, T. Bristol, Shopping with friends and teens' susceptibility to peer influence, J. Retail. 80 (2004) 101–116.

[34] S.P. Ashdown, L. Dunne, A study of automated custom fit: readiness of the technology for the apparel industry, Cloth. Text. Res. J. 24 (2006) 121–136.

[35] S. Ashdown, S. Loker, C. Adelson, Improved Apparel Sizing: Fit and Anthropometric 3-D Scan Data, 2005 (Issue Brief NTC Project: S04-CR01). Available from: http://www.human.cornell.edu/che/fsad/research/ntc.cfm (retrieved 10.02.13).

[36] J.H. Chen-Yu, Y.K. Seock, Adolescents' clothing purchase motivations, information sources, and store selection criteria: a comparison of male/female and impulse/nonimpulse shoppers, Family Consum. Sci. Res. J. 31 (1) (2002) 50–77.
[37] S.B. Kaiser, The Social Psychology of Clothing, third ed., Fairchild, New York, 1997.
[38] D.J.C. Meyer, H. Anderson, Preadolescents and apparel purchasing: conformity to parents and peers in the consumer socialization process, J. Soc. Behav. Personal. 15 (2) (2000) 243–257.
[39] I.J. Grant, G.R. Stephen, Buying behavior of "tweenage" girls and key social communicating factors influencing their purchasing of fashion clothing, J. Fashion Marketing Manag. 9 (4) (2005) 450–467.
[40] J.B. Eicher, S. Baizerman, J. Michelman, "Adolescent dress," in: Book Dress and Identity New York (Fairchild), pp. 122.
[41] F.P. Hughes, L.D. Noppe, I.C. Noppe, Human Development Across the Life Span, sixth ed., West Pub. Co, St. Paul, 1985.
[42] H.R. Hopkins, Adolescence: The Transitional Years, Academic Press, New York, 1983.
[43] T.P. Gulotta, Early adolescence, alienation, and education, Theory into Prac. 22 (2) (1983) 151–154.
[44] B. Ryan, Feminism and the Women's Movement: Dynamics of Change in Social Movement Ideology and Activism, Routledge, New York, 1992, pp. 65–69.
[45] B. Ryan, Feminism and the Women's Movement: Dynamics of Change in Social Movement Ideology and Activism, Routledge, New York, 1992, pp. 71–73.
[46] G.B. Fisher, Pantaloons and Power: A Nineteenth Century Dress Reform in the United States, Kent State University Press, London, 2001, pp. 22–25.
[47] J. Laver, Clothes, Burke, London, 1952, p. 3.
[48] M.E. Roach, J.B. Eicher, Dress, Adornment and the Social Order, John and Wiley & Sons, New York, 1965.
[49] M.R. Roach-Higgins, J.B. Eicher, K.P. Johnson (Eds.), Dress and Identity, Fairchild, New York, 1995, p. 511.
[50] A.M. González, L. Bovone, Identities Through Fashion, Berg, London, 2012.
[51] H. Woo, B. Jin, Culture doesn't matter? the impact of apparel companies' corporate social responsibility practices on brand equity, Cloth. Text. Res. J. 34 (January 2016) 20–36 first published on October 5, 2001.
[52] M. Lewis, C. Dyer, J. Moran, Parental and peer influences on the clothing purchases of female adolescent consumers as a function of discretionary income, J. Family Consum. Sci. 87 (1) (1995) 15–20.
[53] B. Shi, P. Xie, Peer group influence on urban preadolescents' attitudes toward material possessions: social status benefits of material possessions, J. Consum. Affairs 1 (2013) 46.
[54] P. Ward, Consumer socialization, J. Consum. Res. (September 1974) 1–16.
[55] M.R. Solomon, N.J. Rabolt, Consumer Behaviour, Prentice Hall, 2004, pp. 19–21.
[56] M.K. Brock, P.V. Ulrich, L.J. Connell, Exploring the apparel needs and preferences of tween girls and their mothers, Cloth. Text. Res. J. 28 (2) (1995) 95–111.
[57] How to Market to Tweens: Data, Spending Habits+dos and Don'ts to Reach This Fickle Age Group, June 2008. Available from: https://www.marketingsherpa.com/article/how-to/data-spending-habits-dos-donts (retrieved 01.01.16).
[58] D.B. Ryan. How Does Peer Pressure Influence Teen Purchasing Choices? Available from: http://motherhood.modernmom.com/peer-pressure-influence-teen-purchasing-choices-8191.html (retrieved 30.12.15).
[59] E. Drake-Bridges, F. Burgess, Personal preferences of tween shoppers, J. Fashion Marketing Manag. 14 (4) (2010) 624–633.

[60] Teen Spending Trends in 2015, Entrepreneurs. Available from: http://www.businessopportunities.biz/2015/04/15/teens-spending-trends-2015/ (retrieved 24.12.15).
[61] A. Borges, J.C. Chebat, B.J. Babin, Does a companion always enhance the shopping experience? J. Retail. Consum. Serv. 17 (2010) 294–299.
[62] D.H. Demo, The self-concept over time: research issues and directions, Annu. Rev. Sociol. 18 (1990) 303–326.
[63] M. Deutsch, H.B. Gerard, A study of normative and informational social influences upon individual judgment, J. Abnorm. Soc. Psychol. 51 (1995) 629–636.
[64] C. Knorr, Is social media giving your teen a negative body image? Parenting editor. E.M.
[65] J.M. Connolly, V. Slaughter, L. Mealey, The development of preferences for specific body shapes, J. Sex Res. 41 (2004) 5–15.
[66] M.L. Fisher, M. Voracek, The shape of beauty: determinants of female physical attractiveness, J. Cosmetol. Dermatol. 5 (2006) 190–194.
[67] Y. Li, X. Hu, W. Ma, J. Wu, G. Ma, Body image perceptions among Chinese children and adolescents, Body Image 2 (2005) 91–103.
[68] A. Mastamet-Mason, H.M. de Klerk, S. Ashdown, Sizing and fit research at grassroots level. A methodology for the identification of unique body shapes in African developing countries, J. Family Ecol. Consum. Sci. 36 (2008) 9–21.
[69] R. Otieno, C. Fairhurst, The development of new clothing size charts for female Kenyan children. Part II: size-range categorization and presentation of final size charts, J. Text. Inst. 91 (2000) 153–162.
[70] C.R. Jasper, Consumer preferences for size description systems of men's and women's apparel, J. Consum. Affairs 29 (2) (1995) 429–441.
[71] E.J. North, R.B. de Vos, T. Kotzé, The importance of apparel product attributes for female buyers, J. Family Ecol. Consum. Sci. 31 (2003).
[72] F.C. Waksler, Studying the Social Worlds of Children Sociological Readings, Routledge, 1991.
[73] L. Martens, D. Southerton, et al., Bringing children (and parents) into the sociology of consumption towards a theoretical and empirical agenda, J. Consum. Cult. 4 (2004) 155–182.
[74] H. Peterson, Generation Z is a Complete Nightmare for Retailers, June 27, 2014. Available from: http.www.businessinsider.my/generation-z-is-retailers-nightmare-2014-6/#6btxD8Rt2EryJebH (retrieved 25.12.15).
[75] The Teen Apparel Market: Bouncing Back. Supply Chain Insights. Available from: http://www.scribd.com/doc/38580320/Teen-Apparel-Market (retrieved 20.12.15).
[76] Teens Slowly Increase Online Shopping. Market Research World. Available from: http://www.marketresearchworld.net/index.php?option=com_content&task=view&id=4031&Itemid=77 (retrieved 12.06.15).
[77] Mintel International Group Ltd, Children's Wear Retailing – UK, May 2003.
[78] Available from: https://www.commonsensemedia.org/blog/is-social-media-giving-your-teen-a-negative-body-image (retrieved 30.12.15).
[79] Insight into the Children's apparel market, in: Textile Consumer, vol. 32, Cotton Incorporated, New York, 2004.
[80] Global Luxury Retailing, June 30, 2009.
[81] Business Wire, High Growth Forecasted in Global Children's Wear Market, 2007.
[82] S.J.A. Harper, P.J. Dewar, B.A. Diack, The purchase of children's clothing – who has the upper hand? J. Fashion Marketing Manag. 7 (2003) 196–206.
[83] J. Seides, Understanding the Teen Market and How to Maximize Your Swag, February 20, 2015. Available from: http://www.business2community.com/marketing/understanding-teen-market-maximize-swag-01163996#axOTxcbD64OBvCrf.99 (retrieved 01.01.16).

[84] D. Cook, The Commodification of Childhood: The Children's Clothing Industry and the Rise of the Child Consumer, Duke University Press, Durham, 2004.
[85] T.P. Gullota, G.R. Adams, C. Markstrom, The Adolescent Experience, fourth ed., Academic Press, San Diego, CA, 1999.
[86] A.R. Kamaruddin, S. Mokhlis, Consumer socialization, social structural factors and decision-making styles: a case study of adolescents in Malaysia, Int. J. Consum. Stud. 27 (2003) 145–146.
[87] K. Anderson, Preadolescent clothing: an ever expanding challenge and an untapped opportunity, in: [TC]^2 Bi-weekly Technology Communicator, 2006.
[88] L. Craik. 'Children's fashion: small people, big business.' How designer childrenswear grew up to become a billion dollar industry (retrieved 01.02.16).
[89] H. Peterson, Everything You Need to Know About How Teens are Spending Money, What They like, and Where They Shop, April 2014. Available from: http://www.businessinsider.my/how-teens-are-spending-money-2014-4/#Y6YLAfqZ6xBb41bw.99L (retrieved 25.12.15).
[90] A. Kingston, They are Smarter Than Boomers, and Way More Ambitious Than the Millennials, July 15th 2014. Available from: http://www.macleans.ca/society/life/get-ready-for-generation-z/ (retrieved 25.12.15).
[91] K. Peck, Five Things You Need to Know to Market Children Swear to Moms, December 22, 2014. Available from: http://magicblog.magiconline.com/content/5-ways-marketing-to-moms-2015 (retrieved 01.01.16).
[92] J.L. Kestler, Intergenerational Fashion Influences Mother/daughter Relationships and Fashion Involvement, Fashion Leadership, Opinion Leadership and Information Seeking from One Another. Thesis. Ohio University, June, 2010. N.p.: n.p., n.d. Print.

Size and fit of clothing for children and teenagers

2

2.1 Introduction: clothing sizing and fit

According to Abraham Maslow [1] in his theory of Maslow's hierarchy of needs, clothing is considered one of the physiological needs, which are the basic needs for humans. In addition, clothing plays many important roles for health and well-being especially in children's learning and development [2]. Moreover, the roles of clothing include protection against harsh climate and environment, modesty in social and cultural situations, and a form of adornment and an expression of personal taste or style [3]. The most crucial role of clothing is during the early years of human life, which are collectively known as childhood [4].

Children's clothes need to be functional in order to accommodate growth, provide comfort, promote safety, and foster a sense of independence [2,5]. Little research has been conducted to prove that clothes for children need to be functional. For example, Norum [6] mentioned that regardless of how attractive an item is, how easy the garment is to care for, or what value the clothing is to be, it is of no benefit if the garment does not fit the child. In addition, Gautam [4] stated that comfort features were considered as garment size, garment not too tight, and seasonally appropriate clothing with shoes and stockings of correct size. Several recent studies [7–10] have suggested that better fitting and more comfortable clothing for obese and overweight children can increase participation in activities from a social and psychological perspective.

Since the beginning of ready-to-wear garments (RTW) in the mid-1700s, there has been a need for garment size designations that divide body shapes and sizes into different categories [11,12]. This requires a systematic and scientific system to measure and classify the human bodies, which is the purpose of developing a sizing system for garments. Furthermore, a good sizing system that results in a good-fit garment must be based on anthropometric data [13–16]. In another study, Bye et al. [16] mentioned that measuring the human body is a precursor to develop a garment to fit the body. In addition, Gupta [17] mentioned that there are two critical elements that determine the accuracy of a garment system: the anthropometric data and the approach or method used for data analysis and interpretation. Hence, fit and size have been thought of as key factors in garment sizing, which in turn is regarded as the main attribute of garment quality in terms of customer satisfaction [11,15,19,20].

Any sizing system should be developed considering these elements: the type of garment to be produced, how many and what body dimensions should be used to classify the population, and how to label the sizes [21,22]. The elements of a sizing system comprise key (control) dimensions and secondary dimensions, size range, size

Clothing for Children and Teenagers. http://dx.doi.org/10.1016/B978-0-08-100226-1.00002-3
Copyright © 2016 Elsevier Ltd. All rights reserved.

interval, size roll, and size designation [23]. Size designations are also known as the size label [24].

The development of a good sizing system depends heavily on the size range and size intervals. Size intervals are the values that are used to divide the range of a specific sample group. For example, if the size ranges of height for a sample group are 114–152 cm, that makes a range of 48 cm. The size interval is set at 6 cm; therefore, the size roll is estimated to be eight sizes. If the size interval is 8 cm, the size roll is reduced to six sizes and so forth. Therefore, the size intervals play a role in how many sizes in the roll and how many people can match those particular proposed sizes. As the human body is not proportionate and is not the same from one another person, it seems more difficult to cover so many people using standard size intervals. After determining the size range, a decision is made as to how many sizes should be formed and, finally, how many garments should be made in each size. The size interval plays a role in the development of an efficient sizing system because the interval produces the sizes for the whole system [23].

Determination of the size range is based upon whether to use the extreme values or a percentile value. If extreme values are used the range may be too wide; in that case the efficiency of the sizing system will be poor because there will be sizes that fit few people and the coverage rate for the sizing system may be low [23]. The decision of whether or not to use extreme values depends also on the population measured and whether or not there are many people at the extremes.

If there are many people at the extremes, then the accommodation rate for a sizing system that incorporates the extremes may be high. In contrast, if there are only a few people at the extremes and the distance between the measurements of the last person at the quartile and those at the extreme values is wide, then a sizing system that incorporates these few outliers might not be able to accommodate many people as a whole. In sum, although the accommodation rate is high when using the extreme values, the number of resulting sizes will be high too [23].

The size tables contain the size roll, size ranges, and size intervals. After developing the size tables, three conditions are used to evaluate the efficiency of the sizing system: accommodation rate, goodness of fit, and number of sizes. The accommodation rate (or coverage rate) means the percentage of the total population that is covered by the sizing system. A good sizing system is expected to have a high coverage rate. For example, if the accommodation rate for a cluster group of females 7–12 years old is 95%, this indicates that 95% of the female population in that age range will fit into one of the sizes in the size range.

For RTW, having an acceptably good fit means that a manufacturer must aim to cover at least 80% of the surveyed sizes, the 10% of each of the extremes not being catered to [24]. An accommodation of greater than 80% would be ideal but would place excessive requirements on the size roll making it impracticable. In addition, data at the extremes tend to be unreliable as the extremes do not represent a large sample size. Sizing systems typically accommodate between 65% and 80% [23].

Aggregate loss looks at the goodness of fit of garments created based on the sizing system [25–27]. Each size developed for a sizing system is called an assigned size, whereas the size of a given human being is called the actual size. Thus, the goodness of fit for a size can be evaluated by measuring the distance between real and assigned sizes using aggregate loss, a Euclidean mathematical technique for measuring the distance between two points. An efficient sizing system should have a low aggregate loss, meaning that the assigned sizes are very close to the actual sizes and therefore the fit of the clothing is anticipated to be good [28,29].

2.2 Physical growth of children and teenagers

Human life is divided into two stages: child and adult. The legal definition of a child is a person between 2 years and 19 years of age [23]. Biologically, a child is anyone in the developmental stage of childhood, between infancy and adulthood [30]. The United Nations (UN) Convention on the Rights of the Child defines a child as every human being below the age of 18 years [31]. For research that is focusing on children, the term *children* is appropriate, according to the previous definitions, to describe the target population as between 7 and 17 years old. These children belong to the school years, which include elementary or primary schooling, which starts at age 6 or 7 and ends at age 12. For secondary or middle and high schooling, which starts at age 13 and is completed at age 17 or 18.

During childhood and adolescence, physical activity is promoted as a key component of energy balance and as a lifelong positive health behavior [32]. As a result, children are expected to wear comfortable clothing for various activities in order to be physically, socially, and mentally active and fit [33]. On the other hand, during this time children grow quickly, in different ways and at different rates; therefore clothing size is significant in giving them the right fit that allows room for growth [34,35].

In the UN Convention on the Rights of the Child, Article 27 states that children have the right to a standard of living that is good enough to meet their physical and mental needs with regard to food, clothing, and housing [36]. Kang et al. [5] added that when a garment is properly styled according to the right size, it reflects the children's growth and also does not disturb body movements. Therefore, it is also important to wear good-fit clothes especially those that have the greatest potential for growth and body development [37,38].

Infancy and early childhood are both periods of rapid physical growth; these two stages are from birth to 12 months old and then between about 9 and 15 years of age [39,40]. It is therefore to be expected that the fit for clothing is important during these two stages of rapid physical growth. In addition, these rapid physical changes have a serious effect on the fit of clothing if it is not manufactured according to the body shapes and sizes of the targeted population [41–43]. Moreover, figure types play a decisive role in a garment system and contribute to the issue of fit [25].

Studies have shown that fit is significantly associated with customer satisfaction in shopping for children's clothing [6,17]. The same issue was highlighted by Otieno [44], which revealed that parents were dissatisfied with garment fit. She also claimed that customer satisfaction regarding children's clothing fit is unique from other measures of customer satisfaction because the needs of both the parents and the child have to be taken into consideration. Furthermore, in another study, the demand of different

houses is especially true for those parents whose children are an irregular size; sizing then becomes a great concern [45]. Furthermore, ensuring clothes are comfortable and fit for children with different degrees of physical, sociopsychological, and cognitive development can be a challenge to all parents and caregivers [46].

On the other hand, the challenge for manufacturers and retailers lies in understanding the many different factors that can have an impact on the satisfaction of clothing fit, such as functionality, aesthetic appeal, body image, peer influence, and others [47,48]. The issues of body image, popular trends, body shape, and size on clothing selection and fashion consumption are important in today's society, especially among image-conscious females [49–51].

2.3 Factors influencing the growth of children's and teenagers' clothing sizes

Growth, maturation, and development are major processes that children go through from birth to adulthood [52,53]. Growth refers to increase in the size of the body or specific parts of the body [34]. Development refers to the stages of increment of these traits throughout the childhood lifespan while maturation refers to the stage of complete development [33]. During the development process, there are important changes in body size and proportions.

According to Bogin [54], there are three broad stages of children's development: early childhood, middle childhood, and adolescence. In addition, referring to Fig. 2.1, early childhood is considered the infancy stage, middle childhood consists of childhood and puberty, and adolescence is the final stage [54,55].

At puberty, rapid growth occurs in both weight and height, which changes the body size. Girls usually start and finish puberty and adolescence earlier than boys, that is girls from about 10 to 12 years old and boys from about 11 to 14 years old [56]. It is also said that the characteristic physical differences between boys and girls develop during puberty [57]. Moreover, this development is due to the changes in hormones produced by the body [58]. It is also observed that the most important growth spurt

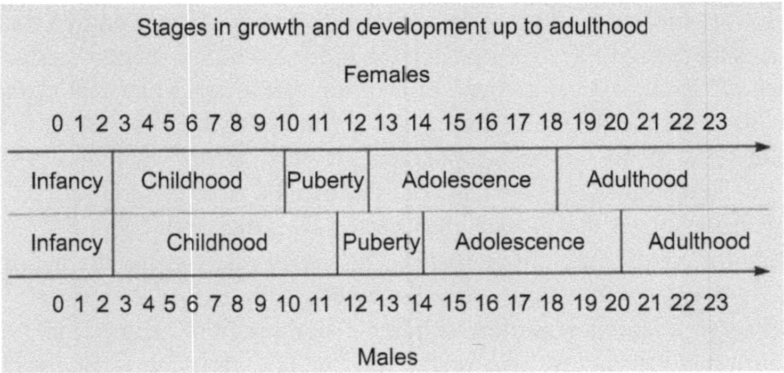

Figure 2.1 Stages of growth and development from childhood to adulthood [55].

occurs during puberty as a part of maturation, which is hormonally driven [40]. Hence, environmental factors such as nutrition, lifestyle, and health appear to be the main causes for differences in growth and maturation between ethnic and social groups [58]. In conclusion it could be said that differences in final height are mainly due to prepubertal growth [59].

The growth spurt produces a rapid increase in both weight and height but it manifests differently in boys and girls [35]. Ahead of this growth spurt, there are no important differences between boys and girls in weight and height. However, after that the differences are obvious from Fig. 2.2, which shows that the differentiation of height occurs between males and female at puberty. Girls reach puberty around age 10, and at the age of 12 they are on average taller than the boys. On the other hand, males reach puberty around age 12, and then by the adolescent stage, age 14 and above, they overtake females in height [24].

The last stage before adulthood is adolescence (13–17 years old), which is considered the most crucial years for children before they begin their lives as adults [60]. Moreover, sexual maturation is also one of the most significant developments during this time [61]. However, sexual development occurs at different ages in different populations so comparing children of the same chronological age but different puberty stage can give different results that lead to different body size and shapes [62]. In addition, growth is said to be completed between the ages of 16 and 18, at which time the growing ends of bones fuse [63].

Body height is clearly of primary importance in fitting clothes for growing children and is recognized as such by ISO TC 133 [64,65]. ISO TC 133 is the recognized ISO technical Committee on Sizing Systems and Designations for Clothes. Moreover, children have their own body structure, which is not simply a smaller version of adults. Often children's sizing systems are developed similarly to adult sizing systems, which can cause a lot of fit problems since the frequency of growth among children varies significantly [25,66]. Another aspect of children's sizing systems is that they are often built around age-related growth spurts for the average child [67]. Because few children are precisely average in all body dimensions, it has been very difficult for parents to find their child's size. In other words, age-related sizes may be inappropriate as a basis for a sizing system if a child grows more quickly or slowly than the average child of that age.

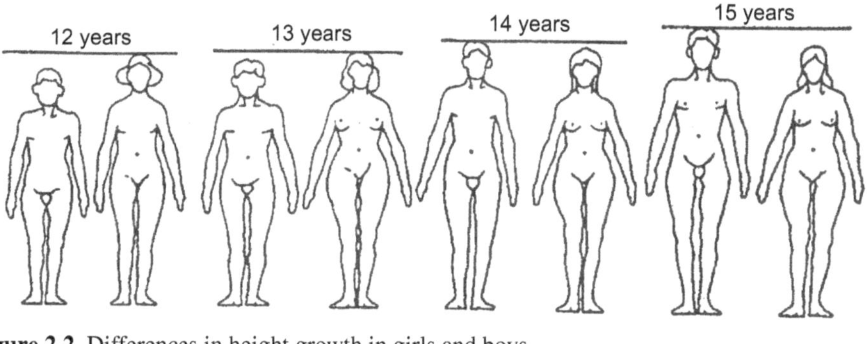

Figure 2.2 Differences in height growth in girls and boys.
Adopted from J.M. Winks, Clothing Sizes: International Standardization, 1997.

2.3.1 Body proportions

The human body is the entire structure consisting of a head, neck, torso, two arms, and two legs. Human vertical body proportion is typically divided into three areas as shown in Fig. 2.3. These divisions are head length, torso length, and leg length. The torso (also sometimes referred to as the trunk) is the anatomical term for the central part of the body from which extend the neck and limbs. The torso includes the thorax and abdomen (see Fig. 2.3).

The physical proportions of the body at birth are very different from those of the adult. Some body parts grow more than others do during development to reach the final adult proportions. Fig. 2.4 shows the relative size of body parts at different ages. However, the age of the different phases shown in the diagram cannot be identified as it is not given in the literature. Changes in the size and shape of the body are caused by different segments growing at different times and different rates [68]. From birth to toddler, the child has a proportionally bigger head; as the child grows older, the head becomes proportionally smaller. The proportions of torso and head differ from one country to another; it has been reported that such differences are due to both genetic and environmental factors [69].

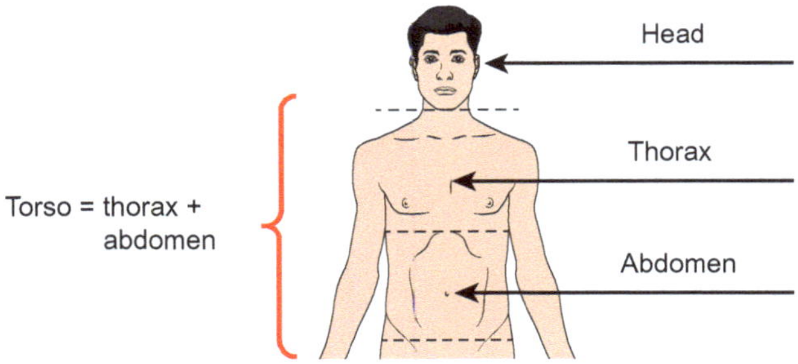

Figure 2.3 The torso section.

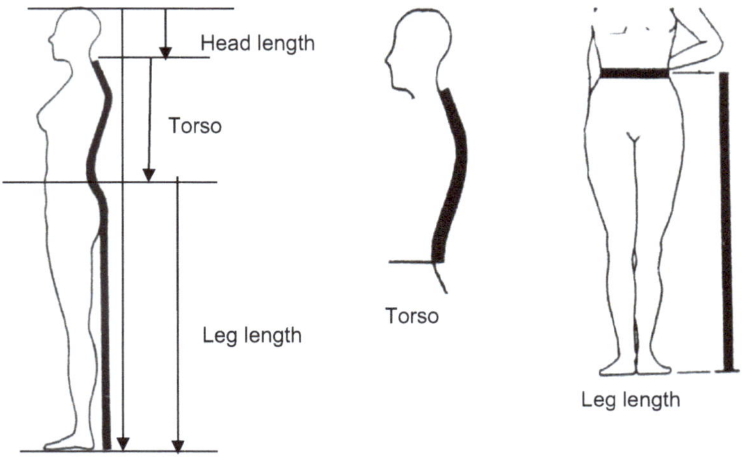

Figure 2.4 The vertical proportion of human body.

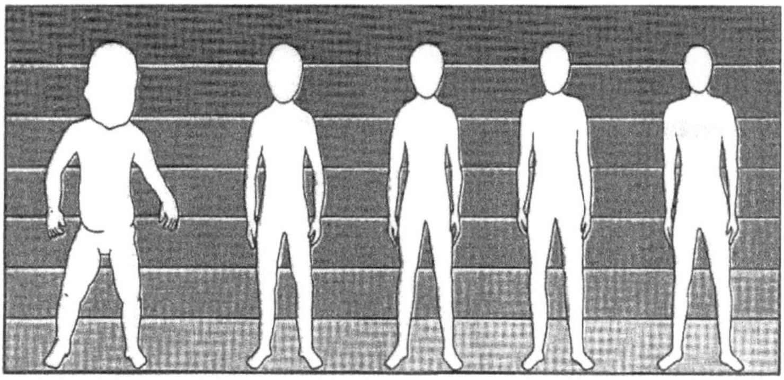

Figure 2.5 Changes in proportion from birth to adult.
Adopted from The International Association of Athletics Federations (IAAF), Introduction to Coaching theory: growth and development, in: IAAF Coaching Theory Manual, 2009.

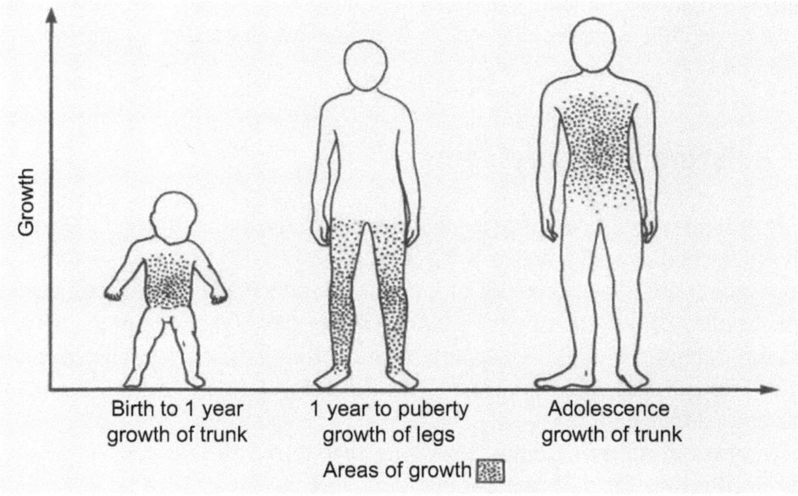

Figure 2.6 Amount of growth from infancy to adolescence.
Adopted from The International Association of Athletics Federations (IAAF), Introduction to Coaching theory: growth and development, in: IAAF Coaching Theory Manual, 2009.

In addition to proportional differences, the area of growth differs during different phases of life. Fig. 2.5 illustrates the different areas of body growth at different life stages. During infancy, the most growth is in the torso; during childhood (from 1 year to puberty) the concentration of growth is in leg length. During adolescence, the growth is again in the trunk [55] (Fig. 2.6).

Bogin et al. [59] said that African American and Australian individuals have the longest legs, due probably to genetic factors. On the other hand, people who live in a healthy environment also proved to have longer legs. In another study, Leicth [70] showed that quality of nutritional and health environment has an impact on body proportions. Children with poor nutrition are underdeveloped, with normal or nearly normal head size but a moderately retarded trunk and relatively short legs. This finding is supported by another study of Mexican American children (2–17 years) in which children from better socioeconomic conditions (ie, better-off families with better food supplies and nutrition) had significantly longer legs but equal trunk length compared with boys and girls from poorer families [71].

Dangour [72] points out that Amerindian children living in Guyana showed tremendous differences in leg length when compared based on their socioeconomic level. He also found that those with better living conditions were taller solely due to leg length, as shown by the fact that there were no differences in the sitting height between the two tribes. This view is supported by Norgan [73] who writes that leg length, head shape, and other aspects of body size are determined by a complex interaction between environmental and genetic factors.

Moreover, in a study of Mayan children (5–12 years), Bogin et al. [59] revealed that body proportions, especially leg length, are a sensitive indicator of the quality of the environment. On the other hand, there is evidence that different socioeconomic or genetic factors affect leg length more as compared to trunk growth. As a conclusion, previous research has proven that stature increases are due to leg length more than to trunk length [74].

2.4 Existing sizing systems

A sizing system is a table of numbers that represents the value of key dimensions used to classify the bodies encountered in the population [75–77]. A sizing system based on anthropometric data (measurement of humans) is known as an anthropometric sizing system. Studies of anthropometric sizing systems have been comprehensively conducted throughout the world for apparel manufacturing efficiency. The structure of a sizing system is based on the division of the target population into subgroups that have similar body dimensions.

Researchers in apparel studies have long recognized the need for consistent and standardized sizing for different populations. Such a system aims at providing each individual a good fit of garment. Moreover, people in each size group are assumed to have the same body shape characteristics and therefore can share the same garment size [25]. Consequently, Petrova [23] stated that the aim of any sizing system is to find the optimum number of size groups that will describe as many shapes and sizes encountered in the population as possible. Another aim is to accommodate as many individuals in each size as possible with a well-fitted garment—at the same time not requiring so many different sizes that it becomes economically unfeasible for the manufacturer [23].

In addition, Salusso et al. [28] mentioned that an efficient sizing system must be based on a classification of the data that accommodates most of the population but with a minimum number of sizes. Moreover, being economically unfeasible, too many sizes to choose from will create confusion for customers and lead to frustration in finding the right size [29,78,79]. Consequently, a sizing system should also not have too few sizes because this results in size groups with dissimilar body dimensions and thus the fit of the garments is not good [21,80]. Ideally, the number of sizes should be neither too few nor too many because either extreme gives dissatisfaction to customers causing reduced profit to the manufacturer [23,81].

Scientists have developed sizing systems for the populations of different countries for one main purpose: to ensure that clothes fit as many people as possible [82,83]. According to Beazley [75], a sizing system also helps customers in identifying the garments that will provide them with a reasonable fit. Moreover, previous researchers have then recommended that a sizing system is derived from anthropometric data using statistical modeling or statistical techniques that should be able to cover the wide variations between individuals [13,75,84].

It is also known that sizing systems are divided into two types: body sizing systems and the garment sizing systems. A body sizing system is a size table based on skin measurements that divide the population into homogenous subgroups, while a garment sizing system is a chart in which ease and allowance are added to the skin measurements to give shape to the garment [23]. This study is only focusing on body sizing systems.

Table 2.1 illustrates the historical development of the anthropometric survey from 1921 to the present day [85]. Anthropometric surveys are carried out extensively dated

Table 2.1 Historical development of sizing systems using anthropometric data

Anthropometric survey for the development of sizing system	Year conducted	Sample size
First American anthropometrics studies were conducted during the end of World War II for males.	1921	100,000 males
A US size survey of some boys and girls was conducted manually on a national basis.	1937–1941	147,000 children
American women size survey using manual measurement.	1939–1940	10,041 females
France surveyed male subjects ages 22–64.	1965–1966	7283 males
Germany published size tables of outerwear for women, men, girls, and boys after conducting a survey.	1961–1966	10,000 males
First Australian survey of women.	1969	11,455 females
Chinese national size survey measuring men, women, and children from 10 different provinces.	1987	14,000 people
Japanese survey of people ages 7–90 using a 3D body scanner.	1992–1994	34,000 people
Size UK used TC^2 3D body scanner.	1999–2002	11,000 adults
Netherlands used Cyberware 3D body scanner.	2000–2002	2000 adults
Size USA used TC^2 3D body scanner.	2002–2003	10,800 adults
Mexico's survey of men and women used TC^2 3D body scanner.	2004	6000 adults
Size Thailand used TC^2 3D body scanner [165].	2007–2008	12,700 adults

as far back as 80 years ago in most of the developed countries. Nevertheless, the same type of survey is still being carried out today in other developing countries realizing the importance of the data not only for sizing system development but also for the health, medical, ergonomics, nutrition, and other applicable industries. As shown in Table 2.1, the earliest report of an anthropometric survey conducted for a clothing sizing system was in 1921 in the United States; this involved 100,000 men in the US Army [85]. Then, between 1937 and 1941, a national US size survey involving 147,000 boys and girls was completed. Also, between 1939 and 1940, the first study of women's measurements for garment and pattern construction was conducted and published in 1941 involving 10,041 women. Later, in 1947, the British Standard Institution developed standards relating to garment sizing based on anthropometric data of 5000 women [86].

Between 1921 and 1999, countries that have done anthropometric surveys for clothing size systems include the United States, United Kingdom, Germany, the Netherlands, South Korea, France, Japan, and Australia. These surveys involved large numbers of people and were conducted manually, which took quite a long time to survey to update their anthropometric data, as is recommended. Meanwhile, countries that had done the survey earlier conducted second anthropometric surveys to renew the anthropometric data that is anticipated to change. From the year 2000 onward, more countries conducted initial anthropometric surveys using new technology such as a digital camera or three-dimensional (3D) body scanner [86].

The anthropometric survey for sizing system development achieved a new technology breakthrough in the late 1998s: the 3D body scanner [87]. Moreover, this device automatically takes a series of measurements of a subject, reducing the manual effort required to perform anthropometric surveys and making them much easier, less invasive, and faster [87]. In addition, this innovative device takes only a few seconds to extract as many different body measurements from a subject [88]. Several countries have used the 3D body scanner for their national sizing survey, including the United Kingdom (Size UK), US (Size US), Japan (Size Japan), Mexico (Size Mexico), and South Africa. Thailand is the latest country to complete the survey, measuring 12,500 people ranging from 18 to 90 years old at the end of 2008 [89]. Thailand has recently launched Size Thailand for use in its apparel industry and other relevant industries [90].

The next country to conduct a national survey for a clothing sizing system for its adult population will be Malaysia and Saudi Arabia. MySize (Malaysian Sizing system) was launched in May 2009 and will use the same TC^2 3D body scanner equipment that was successfully used on Size UK, Size US, and Size Thailand for the collection of anthropometric data. Saudi Arabia is using the Telmat Symcad II 3D body scanner in a study that was launched in late 2015. Both projects anticipate collecting about 6000–8000 body measurements of male and female adults. All of the previously mentioned surveys involved adults. In addition, Malaysia conducted anthropometric data of 2500 children in 2005 using a manual method. The latest national survey involving children was the Shape GB, conducted by the UK, which was the first large-scale project to measure children across the country using 3D body scanners. Shape GB launched its first project measuring children ages 4–17 years old in 2010 using a 3D body scanner and Phase 2 involves manual measurement of babies and toddlers aged 0–4 years old, the results of which are expected to be ready in autumn 2016.

From the previous literature, it can be seen that many national anthropometric studies have been conducted by scientists to determine the ranges of sizes and shapes that exist in human beings for the development of clothing sizes. Humans vary in body proportions and shape even within the same size categories [83]. Kroemer [91] stated that human beings are created with many varieties of body build. In order to design something for humans that fits them well, one must have knowledge of body dimensions and their relationships. It has conclusively been shown that anthropometric data should be used for the development of a good sizing system [15]. However, the human body is not static; thus it is recommended that anthropometric data be revised at least every 10 years. Since childhood consists of several different stages, this results in more body variations in children than in adults.

Because the human body varies considerably, accurate body size information needs to be collected, analyzed, and understood in order to develop good designs that fit humans well. Measurements must be precise and accurate using the right measuring technique, data storage, and statistical treatments. Different measurement and statistical techniques may result in very different information. An anthropometric study is a type of longitudinal study, capturing the changes in body dimensions over a period. In addition, the longitudinal aspect is important because distinct changes in body dimensions occur at different ages, particularly during childhood and adolescence when body dimensions such as stature change rapidly [52].

Correct and reliable data collected from human measurements provide real information for the clothing industry about the sizes of their customers, which can be applied to sizing systems and size labeling in different ways and in different combinations.

2.5 The importance of clothing fit for children and teenagers

Sizing problems are a universal issue and have been comprehensively studied since the early 1940s [14,92]. One of the most important issues in clothing is the size and fit. In addition, Pechoux [13] points out that size is related to individual body dimensions. Moreover, when RTW clothing is manufactured, it is produced in size sets that are based on composite or average body measurements, but selling clothing entails fitting real bodies into the designated sizes set by the apparel industry [83]. Therefore, providing satisfactory-sized clothing has proven challenging to the industry due to the physiological variations of the human body [79]. Thus this issue highlights the importance of clothing fit for children and teenagers. Another issue for teenagers is to find fashionable clothing that fits well.

According to Petrova [23] there are several possible reasons why clothes might not fit. She mentioned that it is because there is no standardization in apparel sizes or that the sizing system is not based on anthropometric data. It could also be that the size charts are based on obsolete anthropometric data. In addition, errors in the pattern-making process and during garment construction can also lead to clothes that do not fit [18]. Moreover, most sizes are based on the assumption that most bodies are hourglass in outline or always of perfect shape, which is clearly not the case. The problem does not only stop at human variations but also can occur at the processes in cutting room, sewing room, and quality control that also affect sizing issues. As Mason and Ashdown [93] emphasized, the dimensions of the human body are the underpinnings of an effective sizing system. Therefore, sizing problems become crucial if the first criterion, which is to base the sizing system on accurate body measurements, is not fulfilled.

Traditionally, sizing systems for RTW clothing utilize a "base size." Once an apparel company has developed a sizing system, they select a suitable model, known as the "fit model." This person is usually of small average size. Consequently, a prototype garment is created and then tried by the fit model and perfected for good fit. Thus, this becomes the base size [94]. Thereby, this garment is then increased and decreased proportionally to set a scale of patterns for the rest of the size range. According to Ashdown [95], the problem that arises from this method is that the sizes are graded up and down based on one person's body shape, which cannot be done because in reality human body dimensions vary widely.

The practice of size inflation (also known as vanity sizing) has been possible partially because of the lack of conformity to regulated garment sizing standards. Otieno et al. [94] affirms that standardized sizes do not accommodate every size and shape but rather assume the concept of the "average" figure, which is usually between sizes 10–14. This means that the average figure is taken as the standard size where if the size is smaller than the standard size, it is increased and if not, it is decreased according to standard fixed increment.

Some of the evidence that shows children need good-fit garments is highlighted in the following paragraphs. Lee et al. [96] stated that apparel cannot be top quality unless it fits the potential wearers satisfactorily. In addition, some studies showed that children look for fit and size as the most important element when purchasing clothing [97–99]. In contrast, teen market consumers were reported to look for qualities in branded clothing that including good fit, look and style [100,101]. Hence, when teenagers try on clothing in the dressing rooms, decisions to reject or adopt the garments are due to three main factors: fit, styling, and appearance [102–104].

Existing apparel companies have worked toward producing clothing according to the unique set of body measurements to fit their own target market like, for example, designing for the plus-sized tweens and teen's body range instead of using the general type of sizing system [105,106]. It has been acknowledged that it is important to produce clothing according to the body measurements of the target population [107]. Moreover, Tselepis [47] mentioned that of all age groups, adolescents are most likely to spend a lot of time and give special attention to clothing and appearance.

Besides their purchasing power and interest in style, the physical development of the early adolescent (9–13 years old) may result in functional and aesthetic problems with the fit of clothes, which in turn might play a part in expectations of fit [108–110]. In addition, elements of fit and size are considered as two of the important attributes for clothing purchase [95]. Therefore, to ensure that clothing is comfortable and fits children well, their body variations need to be studied. This study of human body dimensions is known as anthropometry and is elaborated in the next chapter.

2.6 Conclusions and future trends

As we have seen from this chapter, sizing system construction has been exhaustively studied for many years in most developed countries for improving fit, and many methods have been tested to improve their accuracy and thus the fit. There is no doubt that more studies will be conducted in different parts of the world as manufacturers realize the importance of a standard size system for clothing markets. Existing apparel companies have worked toward producing clothing according to the unique set of body measurements to fit their own target market like, for example, designing for the plus-sized tweens and teen's body range instead of using the general type of sizing system [105–107]. Today it has been acknowledged that it is important to produce clothing according to the body measurements of the target population [107]. Moreover, Tselepis [47] mentioned that of all age groups, adolescents are most likely to spend a lot of time and give special attention to clothing and appearance.

Besides, the teenage purchasing power and interest in style, the physical development of the early adolescent (9–13 years old) may result in functional and aesthetic problems with the fit of clothes, which in turn might play a part in expectations of fit [108–110]. In addition, elements of fit and size are considered as two of the important attributes for clothing purchase [95]. Therefore, to ensure that clothing is comfortable and fits children well, it is recommended for future clothing industry to include these elements as the specifications of children and teenage design factors:

- The understanding of body sizes and shapes
- The size factors based on local body proportions
- The fit factors based on local body dimensions

This study of human body dimensions is known as anthropometry and is elaborated in the next chapter.

References

[1] A.H. Maslow, A theory of human motivation, Psychol. Rev. 50 (1943) 370–396.

[2] A. Stonehouse, Children's Clothing in Child Care, Australia National Children Accreditation Council Inc. (NCAC), 2008.

[3] C.P.S. Boden, J. Pilcher, T. Edwards, New Consumers? The Social and Cultural Significance of Children's Fashion Consumption, 2004.

[4] S. Gautam, Concept of general consideration of clothing for preschool children (2–4 years) among rural mothers of Palampur tehsil, J. Soc. Sci. 11 (1995) 253–254.

[5] Y. Kang, H.S. Choi, W.H. Do, A study of the apparel sizing of children's wear. An analysis of the size increments utilized in children's wear based on an anthropometric survey, J. Korean Home Econ. Assoc. Engl. Ed. 2 (2001).

[6] P.S. Norum, Consumer satisfaction with shopping experiences and clothing products in the children's wear market, J. Consum. Stud. Home Econ. 1 (1995) 331–347.
[7] L.J. Connell, P. Ulrich, K. Simmons, D. Pascoe, D. Bruner, Apparel Product Development for Plus-sized Tween and Teen Boys, Project No. S06-AC03 (Competency: Management Systems), 2006.
[8] K. Caroll, M. Alexander, V. Spencer, Exercise clothing for children in a weight-management program, J. Fam. Consum. Sci. 99 (2007) 68–72.
[9] T. Connolly, Plus Size Clothing Industry Trends: Does Bigger Mean Bigger Business? Special to Fashion Industry Today.com, 2008.
[10] L.J. Chen, K.R. Fox, A.M. Haase, Body shape dissatisfaction and obesity among Taiwanese adolescents, Asia Pac. J. Clin. Nutr. 17 (2008) 457–460.
[11] A. Beazley, Size and fit. Procedures in undertaking a survey of body measurements, J. Fash. Market. Manag. 2 (1997) 55–85.
[12] C.J. Salusso, Discussant: fit and sizing, in: C. Ladisch (Ed.), Proceedings of the International Textile and Apparel Association. Pasadena, 1995.
[13] B. Pechoux, T.K. Ghosh, Standard sizing system and fit testing applied to women's hosiery, Text. Prog. 32 (2002) 11–33.
[14] M.L. Staples, D.B. DeLury, A system for the sizing of Women's garments, Text. Res. J. 19 (1949) 346–354.
[15] A.M. Mason, H.M. de Klerk, S. Ashdown, Sizing and fit research at grassroots level. A methodology for the identification of unique body shapes in African developing countries, J. Fam. Ecol. Consum. Sci. 36 (2008) 9–21.
[16] R. Otieno, Approaches in researching human measurement: MMU model of utilising anthropometric data to create size charts, EuroMed J. Bus. 3 (2008) 63–82.
[17] E. Bye, K.L. LaBat, M.R. De Long, Analysis of body measurement systems for apparel, Cloth. Text. Res. J. 4 (2) (2006) 66–79.
[18] D. Gupta, Indian Body Dimensions, Tex exchange (TC2), 2008.
[19] L.D. Burns, N.O. Bryant, The Business and Fashion Designing, Manufacturing and Marketing, 2002.
[20] R.M. Laing, E.J. Holland, C.A. Wilson, B.E. Niven, Development of sizing system for protective clothing for the adult male, Ergonomics 42 (1999) 1249–1257.
[21] P. Devarajan, C.L. Istook, Validation of female figure identification technique (FFIT) for apparel software, J. Text. Appar. Technol. Manag. 4 (2004).
[22] B. Newcomb, C. Istook, A case for revision of U.S. Sizing standards, J. Text. Appar. Technol. Manag. 4 (2004).
[23] A. Melchiorre, At What Age? Are School-Children Employed, Married and Taken to Court? 2004.
[24] J.M. Winks, Clothing Sizes: International Standardization, 1997.
[25] M.J. Chung, H.F. Lin, J.J. Mao, J. Wang, The development of sizing systems for Taiwanese elementary-and high-school students, Int. J. Ind. Ergon. 37 (2007) 707–716.
[26] C.E. McCulloch, B. Paal, S.P. Ashdown, An optimization approach to apparel sizing, J. Oper. Res. Soc. 49 (1998) 492–499.
[27] D. Gupta, A mathematical approach for developing a garment sizing system for indian population, in: International Conference on High Performance Textiles & Apparels (HPTEX 2004), 2004.
[28] C.J. Salusso-Deonier, M.R. DeLong, F.B. Martin, K.R. Krohn, A multivariate method of classify in body form variation for sizing women's apparel, Cloth. Text. Res. J. 4 (1985) 38–45.
[29] C.J. Yoon, C.R. Jasper, Consumer preferences for size Description systems of Men's and Women's apparel, J. Consum. Aff. 29 (1995) 429–442.

[30] Dictionary, American Heritage Dictionary, 2007.
[31] Convention on the Rights of the Child, United Nations High commissioner for human rights, 1989.
[32] H.W. Kohl, K.E. Hobbs, Development of physical activity behaviors among children and adolescents, Pediatrics 101 (1998) 549–554.
[33] J. McNeal, Children as consumers, Am. Demogr. 12 (1990) 36–39.
[34] T.J. Eveleth, Worldwide Variation in Human Growth, second ed., Cambridge University Press, Cambridge, 1990.
[35] S. Stinson, Growth variation: biological and cultural factors, in: S. Stinson, B. Bogin, R. Huss-Ashmore, D. O'Rouke (Eds.), Human Biology: an Evolutionary and Biocultural Perspective, 2000, pp. 425–463.
[36] UNICEF, UNICEF. The State of the World's Children 1998, Oxford University Press, New York, 1998.
[37] J.H. Yoon, A study on the apparel sizing for children's upper garment, Korean Soc. Costume 3 (1995) 307–321.
[38] A.S. Ariadurai, T.P.G. Nilusha, M.R. Dissanayake, An anthropometric study on Sri Lankan school children for developing clothing sizes, J. Soc. Sci. 19 (2009) 51–56.
[39] F. Bellisle, Effects of diet on behaviour and cognition in children, Br. J. Nutr. 92 (2004) S227–S232.
[40] J.M. Tanner, Fetus into Man, 1978.
[41] J.A. Roebuck, Anthropometric Methods: Designing to Fit the Human Body, 1995.
[42] S.R. Daniels, Body shape in children and adolescents, J. Pediatr. 152 (2008) 45.
[43] P.T. Slee, Child, Adolescent and Family Development, second ed., Cambridge University Press, New York, 2002.
[44] R. Otieno, The role of garment sizing in creation of customer satisfaction: indications from focus groups responses, J. Fash. Market. Manag. 4 (2000) 325–335.
[45] K. Yeosun, C. Hei-Sun, H.D. Woei, A study of the apparel sizing of children's wear – an analysis of the size increments utilized in children's wear based on an anthropometric survey, J. Korean Home Econ. Assoc. Engl. Ed. 2 (2001) 96–110.
[46] S.J.A. Harper, P.J. Dewar, B.A. Diack, The purchase of children's clothing – who has the upper hand? J. Fash. Market. Manag. 7 (2003) 196–206.
[47] T. Tselepis, H.M. De Klerk, Early adolescent girls' expectations about the fit of clothes: a conceptual framework, J. Fam. Ecol. Consum. Sci. 32 (2004) 83–94.
[48] A. Kiran, A. Riaz, N.H. Malik, Factors affecting change in the clothing patterns of the adolescent girls, Int. J. Agric. Biol. 4 (2002) 377–378.
[49] D. Mellor, M. McCabe, L. Ricciardelli, J. Yeow, N. Daliza, N.F.M. Hapidzal, Sociocultural influences on body dissatisfaction and body change behaviors among Malaysian adolescents, Body Image 6 (2009) 121–128.
[50] J. Cajigas, Strategies, how to Sell Kids' apparel, in: Stitches, Embroidery's Voice and Vision, 2009.
[51] D. Martin, Shops forced to make bigger children's clothes to cope with obesity epidemic, Mail Online (2009).
[52] S. Sigelman, Life Span of Human Development, fourth ed., Thomson Learning, New York, 2006.
[53] C.B. Malina, B.O. Oded, Growth, Maturation, and Physical Activity, Human Kinetics, Champaign, IL, 2004.
[54] B. Bogin, Patterns of Human Growth, second ed., Cambridge University Press, 1999, p. 444.
[55] The International Association of Athletics Federations (IAAF), Introduction to coaching theory: growth and development, in: IAAF Coaching Theory Manual, 2009.

[56] T.P. Gullota, G.R. Adams, C. Markstrom, The Adolescent Experience, fourth ed., Academic Press., San Diego, CA, 1999.
[57] C. Ruff, Variation in human body size and shape, Annu. Rev. Anthropol. 31 (2002) 211–232.
[58] L.A. Proos, Anthropometry in adolescence-secular trends, adoption, ethnic and environmental differences, Horm. Res. 39 (1993) 18–24.
[59] B. Bogin, P. Smith, A.B. Orden, V.M.I. Silva, J. Loucky, Rapid change in height and body proportions of Maya American children, Am. J. Hum. Biol. 14 (2002) 753–761.
[60] S.G. Dudek, Nutrition Essentials for Nursing Practice, fourth ed., Lippincott, Philadelphia, 2001.
[61] M. Voracek, M. Fisher, Shapely centrefolds? Temporal change in body measures: trend analysis, BMJ 325 (2002) 1447–1448.
[62] P.T. Slee, Child, Adolescent and Family Development, second ed., 2003, pp. 320–331.
[63] J. MacGregor, Introduction to the Anatomy and Physiology of Children: A Guide for Students of Nursing, Child Care and Health, 2008, p. 255.
[64] ISO/TR 10652, Standard Sizing Systems for Clothes, 1991.
[65] ISO 3637, Size Designation of Clothes – Women's and Girls' Outerwear Garments, 1977.
[66] M. de Onis, A.W. Onyango, E. Borghi, A. Siyam, C. Nishida, J. Siekmann, Development of a WHO growth reference for school-aged children and adolescents, Bull. W. H. O. 85 (2007) 660–667.
[67] J. Robotham, Children's changing shape, in: Brisbane Times, 2009.
[68] V. Abbassi, Growth and Normal puberty, Pediatrics 102 (1998) 507–511.
[69] R.E. Kail, Children and Their Development, fourth ed., Prentice Hall, 2006.
[70] I. Leich, Growth and health, Br. J. Nutr. 5 (1951) 142–151.
[71] A.R. Frisancho, N. Guilding, S. Tanner, Growth of leg length is reflected in socio economic differences, Acta Med. Auxol. 33 (2001) 47–50.
[72] A.D. Dagour, Growth of upper and lower body segments in Patamona and Wapishana Ameriindian children (cross sectional data), Ann. Hum. Biol. 28 (2001) 649–663.
[73] N.G. Norgan, Body proportion differences, in: S.J. Ulijaszek, F.E. Johnston, M.A. Preece (Eds.), Cambridge Encyclopedia of Human Growth and Developement, 1998, pp. 378–379.
[74] D.J. Gunnell, S.G. Davey, S. Frankel, M. Kempel, T.J. Peters, Socioeconomic and dietary influences on leg length and trunk length in childhood: a reanalysis of the Carnegie (Boyd Orr) survey of diet and health in prewar Britain (1937–1939), Perinat. Epidemiol. 12 (1998) 96–113.
[75] R. Otieno, C. Harrow, G. Lea-Greenwood, The unhappy shopper, a retail experience: exploring fashion, fit and affordability, Int. J. Retail Distrib. Manag. 33 (4) (2005) 298–309.
[76] J.E. Croney, An anthropometric study of young fashion students including factor analysis of body measurements, Man 12 (1977) 448–496.
[77] B. Tsang, C.K. Chan, Taylor, Kin anthropometry study of the physique of disciplined personnel, Int. J. Cloth. Sci. Technol. 12 (2000) 144–160.
[78] E. Goldsberry, S. Shim, N. Reich, The development of body measurement tables for women 55 years and older (Part 1), Cloth. Text. Res. J. 14 (1996) 108–120.
[79] K. Desmarteau, Let the Fit Revolution Begin, 2000.
[80] P. Tryfos, An integer programming approach to the apparel sizing problem, J. Oper. Res. Soc. 37 (1986) 1001–1006.

[81] C.H. Hsu, M.J.J. Wang, Using innovative technology to establish sizing systems, Int. J. Innov. Learn. 2 (2005) 233–245.
[82] K.L. LaBat, Consumer Satisfaction/dissatisfaction with the Fit of Ready-to-wear Clothing (Ph.D. dissertation), University of Minnesota, 1987.
[83] S. Ashdown, S. Loker, Improved Apparel Sizing: Fit and Anthropometric 3D Scan, National Textile Center Annual Report: Nov 2004, 2004.
[84] D. Gupta, Anthropometric data analysis and garment sizing system for Indian population, in: Presented at 86th Textile Institute Conference, Hong Kong, 2008.
[85] W. Yu, Human anthropometrics and sizing systems, in: W. Yu, L. Hunter, J. Fan (Eds.), Clothing Appearance and Fit: Science and Technology, Woodhead Publishing, Cambridge, 2004, pp. 174–177.
[86] J. Fan, W. Yu, L. Hunter, Clothing Appearance and Fit: Science and Technology, Woodhead Publishing Limited, Cambridge, 2004.
[87] K.P. Simmons, C.L. Istook, Body measurement techniques: a comparison of three-dimensional body scanning and physical anthropometric methods for apparel application, J. Fash. Market. Manag. 7 (2003) 306–332.
[88] A. Petrova, S.P. Ashdown, Three-dimensional body scan data analysis: body size and shape dependence of ease values for Pants' fit, Cloth. Text. Res. J. 26 (2008) 227–252.
[89] S. Charoensiriwath, A real-time data monitoring and management system for Thailand's first national sizing survey, in: Presented at PICMET 2008. Portland International Conference on Publication, Portland, 2008.
[90] A. Pornwasin, Nectec develops national body chart, in: The Nation. Bangkok, 2009.
[91] K.H.E. Kroemer, H.J. Kroemer, K.E. Kroemer-Elbert, Engineering Physiology: Bases of Human Factors/Ergonomics, third ed., 1997.
[92] R. O'Brien, W.C. Shelton, Body measurements of American boys and girls for garment and pattern construction, in: U.S. Dept. of Agriculture Miscellaneous Publication 366, U.S. Department of Agriculture, Washington, DC, 1941, p. 141. M. P. 454.
[93] D. Mahadzir, In a fit over sizing, in: Sunday Star. 21st August, 2005. Kuala Lumpur, .
[94] L.J. Anderson, E.L. Brannon, P.V. Ulrich, A.B. Presley, D. Woronka, M. Grasso, S. Gray, Understanding Fitting Preferences of Female Consumers: Development of an Expert System to Enhance Accurate Sizing Selection, National Textile Center Annual Report: I98–A08, November 1999.
[95] S.P. Ashdown, L. Dunne, A study of automated custom fit: readiness of the technology for the apparel industry, Cloth. Text. Res. J. 24 (2006) 121.
[96] Y.T. Lee, Body Dimensions for Apparel, NISTIR 5411, National Institute of Standards and Technology, Gaithersberg, MD, 1994, p. 1.
[97] M. Konstanski, A. Fisher, E. Gullone, Current conceptualisation of body image dissatisfaction: have we got it wrong? J. Child Psychol. Psychiatry 45 (2004) 1317–1325.
[98] Z. Zhang, G. Ching, G. Ging, J. Moody, W. Liu, Investigation of denim wear consumption in China, Res. J. Text. Appar. 3 (1999) 60–64.
[99] M. Grasson, R. Wright, African American mother's needs,search and evaluation of children's clothing, J. Fash. Market. Manag. 2 (1997) 41–55.
[100] I.J. Grant, S.R. Graeme, Buying behaviour of "tweenage" girls and key societal communicating factors influencing their purchasing of fashion clothing, J. Fash. Market. Manag. 9 (2005) 450–467.
[101] P.N. Danziger, What People Buy: Teen and Tween Clothing and Apparel, Teen (Ages 13 to 17) and Tween (Ages 7 to 12) Clothing Is the Lowest Purchase Incidence Category, 2004.

[102] S. Ironico, The active role of children as consumers, Young Consum. 13 (1) (2012) 30–44.
[103] M. Eckman, M.L. Damhorst, S.J. Kadolph, Toward a model of the in-store Purchase decision process: consumer use of criteria for evaluating women's apparel, Cloth. Text. Res. J. 8 (1990) 13–22.
[104] K.L. LaBat, M.R. DeLong, Body cathexis and satisfaction with fit of apparel, Cloth. Text. Res. J. 8 (1990) 43–48.
[105] Y.S.C. Ming, Body shape preference and body satisfaction of Taiwanese and Japanese female college students, Psychiatric Research 133 (2005) 263–271.
[106] C.R. Martins, E. Carraca, P.J. Teixeira, A.M. Silva, E.L.P. Petroski, Prevalence of body shape concerns and associated factors among Brazilian early adolescents, Hum. Mov. 15 (1) (2014) 12–20.
[107] M. Alexander, L.J. Connell, A.B. Presley, Clothing fit preferences of young female adult consumers, Int. J. Cloth. Sci. Technol. 17 (2005) 52–64.
[108] D.C. Jones, T. Vigfusdottir, Y. Lee, Body image and the appearance culture among adolescent girls and boys: an examination of friend conversation, peer criticism, appearance magazines and the internalization of appearance ideals, J. Adoles. Res. 19 (2004) 323.
[109] M. Suzanne, J. Lee, Proximity of clothing to self-scale, Cloth. Text. Res. J. 22 (2004) 161.
[110] D.C. Jones, T.H. Vigfusdottir, Y.S. Lee, Body image and the appearance culture among adolescent girls and boys: a…"Posttraumatic growth among adolescents", J. Adolesc. Res. 19 (2004) 851–856.

Conducting anthropometric surveys of children and teenagers

3

3.1 Introduction: anthropometry

Anthropometry literally is defined as "measurement of humans" [1,2]. In Greek, *anthropos* means "human" and *metron* means measure [3]. Anthropometry also is a study of humans where the study focuses on the measurement of living individuals in order to understand the human physical variations [4]. In addition, according to Yu [5] human anthropometry is the study of people in terms of their physical dimensions, which has classically been performed by the physical anthropologist. In other words, *anthropometry* is the formal name for the techniques used to gather quantitative information about the human body that can be applied in a design context [6]. Anthropometry is also conducted in a systematic way to measure the human body to obtain the descriptors of body size and shape. This information is very important for clothing sizing development, especially for targeted populations of each country.

Human bodies differ greatly in dimensions. The variations come from factors of gender, nationality, and environment, as well as other predictable factors like age [7]. Thus, knowing that people in different countries have different body dimensions and proportions dictates the differences in the sizes of clothing. Adopting sizes from other countries to match local sizes will lead to unfit clothing wear. It is of great importance that national anthropometric surveys are conducted to gather profiles of body shapes and proportions so that the sizes can be determined for fit and comfort especially when the variations are really wide as seen in the population of children. The spread of growth is rapid in children and thus the sizes vary widely as compared to the adult anthropometric data.

3.1.1 Principles in the application of anthropometric data

There are three applicable principles of anthropometric measurements for design purposes that can be classified into different situations [8]:

1. *Design for Extreme Individuals*: this feature is relevant for products that can dictate the minimum and maximum value of the population variable or characteristics. The relevant population characteristics use the 5th and 95th percentiles of the distributions for the relevant design parameters, such as door design, which should fit the tallest person or the shortest person.
2. *Design for Adjustable Range*: this type of product feature is relevant for product services that can be adjusted to the person using it. It still covers the range of 5th to 95th percentile but

Clothing for Children and Teenagers. http://dx.doi.org/10.1016/B978-0-08-100226-1.00003-5
Copyright © 2016 Elsevier Ltd. All rights reserved.

focuses on the relevant population, for example, people who use office chairs or automobile seats that need adjustable sitting and arm heights etc.

3. *Designing for the Average Person*: this type of product feature is relevant to individuals who are average on each of many anthropometric characteristics. This feature uses average values in the design of ready-to-wear clothing where the range of sizes is based on the average of each size.

3.2 Key issues in the body measurement process

This chapter describes all the key issues involving the body measurement process. There are four steps, as shown in Fig. 3.1. Steps 1 and 2 describe the fieldwork and planning for both techniques, which are the manual and computerized. Steps 3 and 4 will be different for each technique. The goal of this stage is to collect body measurements of the sample population manually and digitally and then analyze those using simple statistical methods and data mining techniques. The purpose of this analysis is to understand the body ranges and variations of children in a population and then classify them into meaningful groups for the right sizing system.

The whole process is summarized in the flow chart in Fig. 3.2, showing the processes and activities involved from the beginning to the end. Each step is described in the following sections.

3.2.1 Step 1: fieldwork preparation

Fieldwork preparation is the grounding work before conducting the anthropometric survey. It is more reliable to recruit children and teenagers from school grounds. The preparation for recruitment is more structured when dealing with the Ministry of Education. The planning and fieldwork preparation can be targeted safely and schools are more willing to cooperate, thus using their help is more practical. Furthermore, since this is for educational purposes, dealing with the Ministry of Education will ensure that the research is feasible. Activities include writing letters to the authorities for permission to enter schools, developing the anthropometric protocol, and training the measurers. The details of the preparation are described in the next section.

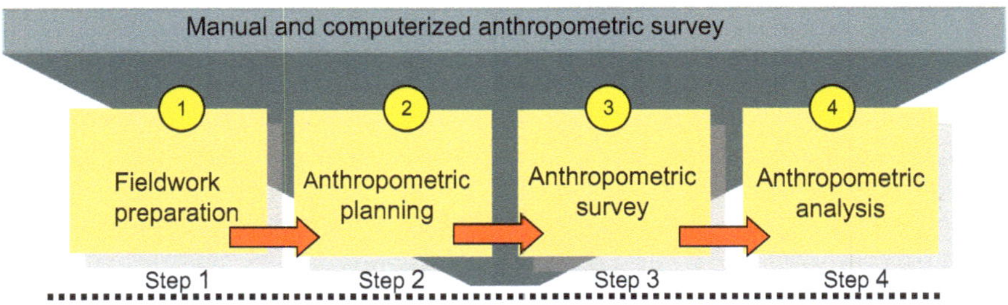

Figure 3.1 Methodology for sizing system development.

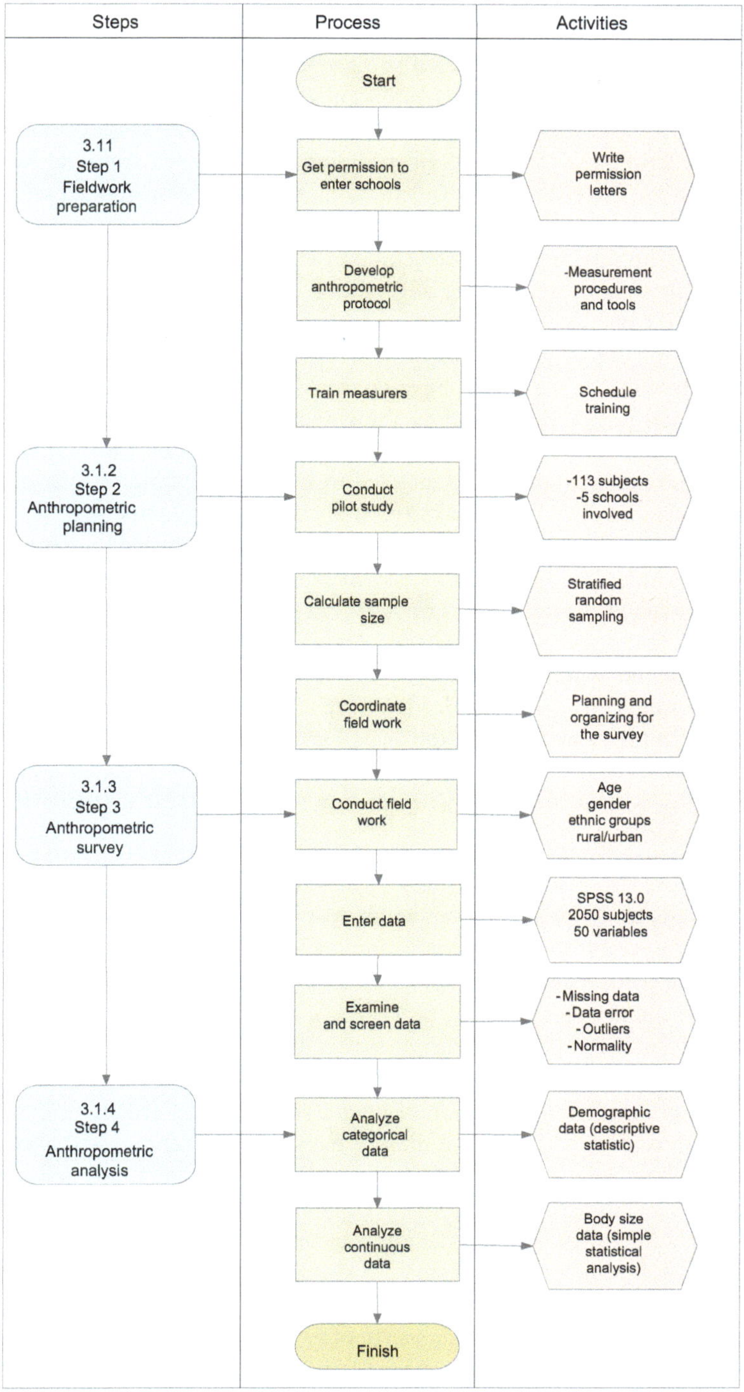

Figure 3.2 Anthropometric process flow chart.

3.2.1.1 Permission from school authorities

The cross-sectional study is conducted at three school levels, namely the elementary, middle, and high school. Collecting children's body data for sampling purposes is usually easier from school as compared to dealing with various other organizations from public recruitment. The formal permission was obtained from the Ministry of Education and the State Education Department. All government-funded schools were targeted first before going to private schools since the deal was made with the Ministry of Education. The sample consists of children in elementary schools covering years 1–6 and middle to high school covering years 1–5. Lists of schools for all the levels can be obtained from the State Education Department. The lists of schools are divided according to rural and urban schools for total coverage.

Once the permission was obtained from the Ministry of Education, the researcher was able to enter any public school by showing the letter of permission. In addition, after obtaining the permission letter to enter the school grounds, formal letters were sent to selected geographical areas targeted to fulfill the sample population requirements. Respondents were informed of the study and were told that their participation was voluntary. Answer slips indicating whether the schools wanted to participate were attached to the letters, and recipients were asked to fax the slip back to the researcher. Some schools called the researchers directly, but most of them faxed back the answer slips within a few days. Schools that responded as willing to participate were then contacted by telephone or by personal visits to the school to ensure that the objectives of the survey were fully understood. After obtaining the school's confirmation to participate through fax or phone calls, the dates for measurement visits were scheduled.

3.2.1.2 Manual and computerized anthropometric measurement training

There was a research team consisting of project manager, technician, research assistant, and measurers to assist the manual process or the scanning process all over the regions. They moved from one region to another region to measure the assigned numbers per region. Thus, before they started the mission, they were given solid anthropometric measurement training.

For the manual anthropometric survey, a training handbook based on ISO 8559/1989 was prepared to explain the terms for each different body dimension. Table 3.1 lists all the 51 body dimensions inclusive of height and weight. Furthermore, these body dimensions were divided into three main groups: vertical length, width length, and girth. In addition, these dimensions were divided into upper and lower body. There are 29 dimensions suitable for upper or whole body marked with asterisk (*). On the other hand, the other 20 dimensions not marked are categorized as lower body dimensions. The diagrams for each body dimension are also shown in the handbook for easy recognition during practices.

The training takes about 30h over a span of five working days. First, the measurers were briefed about the objectives of the anthropometric survey. They were introduced to the topic using a PowerPoint presentation, and the objectives were clearly explained so that consistency and precision were taken seriously. Each trainee was given an

Table 3.1 **List of body dimensions according to ISO 8559/1989**

Length (vertical)	Width (vertical)	Girth (horizontal)
Height		Weight
1. *Under arm length	1. *Shoulder length	1. *Head girth
2. *Scye depth	2. *Shoulder width	2. *Neck girth
3. *Neck shoulder point to breast point	3. *Back width	3. *Neck base girth
4. *Cervical to breast point	4. *Upper arm length	4. *Chest girth
5. *Neck shoulder to waist	5. *Arm length	5. *Bust girth
6. *Cervical to waist (front)	6. *7th cervical to wrist length	6. *Upper arm girth
7. *Cervical to waist (back)	7. *Hand length	7. *Armscye girth
8. *Cervical height (sitting)	8. Foot length	8. *Elbow girth
9. * Trunk length		9. *Wrist girth
10. *Body rise		10. *Hand girth
11. *Cervical to knee hollow		11. Waist girth
12. *Cervical height		12. Hip girth
13. Waist height		13. Thigh girth
14. Outside leg length		14. Mid-thigh girth
15. Waist to hips		15. Knee girth
16. Hip height		16. Lower knee girth
17. Crotch		17. Calf girth
18. Trunk circumference		18. Minimum leg girth
19. Thigh length		19. Ankle girth
20. Inside leg length/crotch height		
21. Knee height		
22. Ankle height		

* Upper body dimensions.
Lower body dimensions.

anthropometric manual with pictures of each body dimension, and each body dimension was thoroughly explained and demonstrated on the real body.

This manual measurement training is good for both anthropometric measurement techniques, manual and computerized, because the training will ensure that the trainers understand the anatomy of the human body. By recognizing and holding all the important parts of the body that need to be measured, they will be very familiar with what needs to be measured, although when using the computerized method they will not touch the body. This training will enhance the knowledge of body anatomy. In addition, measurers also need to conduct the same procedure to check the reliability of the scanning technologies when they use the computerized measurement techniques.

For computerized anthropometric survey, we need to decide what measurements need to be included using the manual method, and also for the digital method, how many body dimensions will be extracted from each subject. For the scanning procedure, manual measurements are completed for certain dimensions like head circumference, height, and weight. The other body dimensions need to be manually measured according to the capabilities of different scanner abilities.

At this point of training, the measurers still need to be trained to understand and become familiar with all the body dimensions as mentioned earlier. With the PowerPoint presentation, the video for the scanning process will also be developed for measurers' training using the scanning methodology and for the subjects to review before the scanning process is conducted. This video is an important document that provides better understanding of the whole process. The actual layout of the scanning procedure is also developed at this point, discussing considerations like the flow of the scanning process, the traffic, the efficiency of each step, and the development of the scanning area including the privacy concerns.

During training sessions, the anthropometric survey team will performed hands-on practice on their partners for at least five days until they were comfortable and familiar with each body dimension. These practices are more crucial if the training is for manual anthropometric survey. Each measurer is given forms, with all the body dimensions listed, to fill in with the measurements. The measurers worked in pairs. The measurement practices continue until they are confident with the technique of measuring and the readings remain consistent throughout the measurement days.

On the other hand, for scanning methodology, the team will be trained not only to do manual measurements but also how to operate the scanning machine. They will practice scanning the participants, get the participants ready to be in the booth and scanned successfully. They will also practice working specifically within the time needed to scan one person, which includes the changing time, and also the manual procedure, in order to calculate the number of participants they can manage in one day.

Lastly, the preparation for manual anthropometric survey includes the ordering of food for the subjects, the preparation for transportation to bring the team, and the tables and chairs for refreshments, and the subjects. As for the computerized anthropometric survey, the preparation will include the planning for the consumable items for the scanning process. Decisions need to be made for transport rental for the measurers to move from one region to another region, transport to bring the mobile 3D scanner and other anthropometric equipment to all the selected regions, the food and drinks arrangement for all the scanning participants, the requirement features for the scan wear production, the acquisitions of headwear and disposable underwear for the scanning participants.

3.2.2 Anthropometric planning

Anthropometric planning is a very important procedure to be conducted for any method of anthropometric survey. For this chapter, the planning is focused for both the techniques, which are the manual and computerized surveys. Anthropometric planning comprised the preliminary study, sample size calculation, and fieldwork coordination. The first purpose of the preliminary study is to test the whole process of measuring, to understand the nature of the survey, and to solve any potential problems before the real anthropometric survey. The second purpose is to get the measurements needed to calculate the sample size for the anthropometric survey.

3.2.2.1 Preliminary study

The preliminary survey was conducted to measure 100 children between 7 and 17 years old using the traditional method, for pilot test purposes. For this survey, letters of research intention were sent to primary and secondary schools. Personal visits and meetings with the school principals were organized at the nearby schools in the selected city after the letters were received.

The principals were then briefed about the importance of the measurement process. The full plan of the preliminary survey was given to the principals of the respective schools. The survey was completed within 10 days, approximately 10 children per day. Each child was asked to change into a tight-fitting garment before they were measured. For each measurement, the time recorded was about 30–40 min. The measurement process was conducted in half-day sessions starting at about 10 am and ending at noon.

The preliminary anthropometric survey was successfully completed at six different schools in the selected districts. From the collected data, the sample size required for the anthropometric survey is calculated.

3.2.2.2 Sample size

In any study it is usually impossible to study the entire population of interest and therefore a researcher usually relies on sampling to obtain a subset of the population [9]. Observations made from the sample survey can be inferred to apply to the entire population of one country, for example, if the state has the largest population of children in this particular country, then the state sampling can be considered for sample size.

Sampling can be taken from five districts based on their regions: north, south, west, central, and east (for example, see Table 3.2). All five districts that are selected for sampling in each region are those situated next to each other, for example. In addition, schools in this state are selected based on their geographical areas: rural and urban. Region and geographical areas are one of the features for sample selection. A school is considered to be in an urban setting if it is situated within the 20-km radius of the town and rural if it is situated beyond this radius. One hundred letters were sent to these districts and of them 29 schools volunteered to participate in the anthropometric survey. The support and understanding from the school management enabled the survey to be conducted efficiently and effectively.

To calculate the sample size, the stratified random sampling technique is applied [10,11]. This method focuses on the groups. In this study, the data collection had four variables: age, gender, ethnic group, and geographical area (rural and urban). This research is based on two groups, which are the age and gender. The proportionate sample size reflects the distributions of real population of age and gender groups in this particular state. The other two parameters, ethnic groups and geographical areas, are

Table 3.2 Population according to districts in one state of a country

Targeted area	Districts of an area	Population ('000)
1	Area A	675.6
2	Area B	1142.5
3	Area C	825.1
4	Area D	198.1
5	Area E	1514.1

selected according to a simple random technique with the targeted numbers of subjects from the calculated proportionate sample size.

Proportionate stratified sampling is used for this research, which means taking the same proportion (sample fraction) from each stratum [10]. For example, there are three groups of students: Group A with 100 people, Group B with 50 people, and Group C with 30 people. These groups are referred to as strata. The sample units are randomly selected from each stratum based on proportion. For example, a proportion of 10% from each group (strata) would mean that 10 people were taken from group A, 5 people from group B, and 3 people from group C. The strata group for this research is based on two groups: age group (7–17 years old) and gender (female and male). The real total numbers of the populations of children in this particular state in this country for each age are given in Table 3.3.

For this research, two steps are taken in order to obtain the number of samples needed for the anthropometric survey. Firstly, the sample size is calculated using the stratified random sampling formula (Eq. [3.1]). The data obtained from the preliminary study are used to calculate the sample size. Then, the numbers of children according to gender and age group are calculated using the proportionate sampling based on the actual number of children in one state. The steps are shown next.

3.2.2.3 Sample size determination

The sample size was calculated using the stratified random sampling formula as shown in Eq. [3.1] [11]:

$$n = \frac{\sum_{i=1}^{l} N_i^2 \sigma_i^2 / a_i}{N^2 D + \sum_{i=1}^{l} N_i^2 \sigma_i^2} \qquad [3.1]$$

where, N = total number of children in one state; X_i = input for sample i; Y_i = output for sample i; x = mean value of input data; y = mean value of output data.

Table 3.3 Number of samples in each group

Age/years	Number of students
7	91,123
8	86,500
9	84,800
10	87,627
11	84,598
12	81,806
13	69,540
14	66,300
15	62,750
16	56,121
17	49,871

State Education Department.

The first process is to calculate the sample size for the survey. The body dimensions that were used to calculate the sample size are height, chest girth, and hip girth. After deciding the total number of sample size, then the sample size for each of the strata parameters are calculated. The total sample size can be chosen from three different variables. Then, the total number of subjects for each of the age ranges and genders are calculated based on the proportionate method formula.

First step:

$$\bar{y}_{\text{st}} = \frac{1}{N}\left[N_1\bar{y}_1 + N_2\bar{y}_2 + \cdots + N_1\bar{y}_1\right] \qquad [3.2]$$

where, N = total population age 7–17; N_1 = total population age 7–12; N_2 = total population age 13–17; $\bar{y}$ = mean of variables for each age group.

Second step:

$$D = \left(\frac{0.01 \times \bar{y}_{\text{st}}}{2.326}\right)^2 \qquad [3.3]$$

Third step:

Calculate the sample size using stratified random sampling as in Eq. [3.1].

The age range for this study is between 7 and 17 years old. A sample population of ages 7–12 are gathered from elementary schools and ages 13–17 are gathered from middle and high schools in one state from one particular country. The whole total population of children in this state within the range of 7–17 years old is 823,071[N]. The number of each age is tabulated in Table 3.3. Each age group is a stratum and is calculated using the proportionate method according to the ratio of the real population. Each stratum age [h] is given by:

$$n_h = (N_h/N)*n \qquad [3.4]$$

where, n_h is the sample size for stratum h; N_h is the population size for stratum h; N is total population size; and n is the sample size.

For this study, both males and females were included for each age, ethnic group, and geographical area. The male-to-female ratio in the real population in one state is noted as 51% male and 49% female. For each age total, the sample is divided into the corresponding ratio (n) of male and female.

$$n_{\text{m}} = N_h*n \qquad [3.5]$$

$$n_{\text{f}} = N_h*n \qquad [3.6]$$

where, n_{m} is the sample size for stratum male and n_{f} is the sample size for stratum female; N_{h} is the population size for stratum h (age); and n is total sample size.

3.3 Manual anthropometric surveys of children and teenagers

Manual anthropometric survey is also known as the traditional anthropometric method. This method uses the equipment that is totally handled manually by measurers that are being trained. The tools can be seen in Figs. 3.3 and 3.4. The manual method of measuring uses tools that are far cheaper as compared to the computerized method. However, the tools need to be handled by measurers who are already familiar and expert in taking the measurements accurately.

Both manual and computerized anthropometric survey research need to be carefully planned since the subject matter is children and teenagers. Additional planning should be considered as the subjects are more challenging to measure and also the children need to be acquired from a convenient sampling place like schools where the permission can be obtained from a governed body, which makes the survey run more smoothly. A manual anthropometric survey of children is different from an adult manual survey because children need more attention and also because the researchers need to obtain consent from the parents. When adults are being measured, the only permission that is needed is from the individual themselves, but when you measure children, you need to get consent from parents and also seek approval from the children if they are willing to be measured or not. Therefore, many delicate and small issues need to be reflected upon when conducting anthropometric surveys for children. The methodology and challenges are described in the next few subtopics.

3.3.1 Measurement tools

To conduct an anthropometric survey for the clothing industry, the ISO standard (ISO 8559:1989, garment construction and anthropometric surveys–body dimension) is adopted as the guideline for taking body measurements.

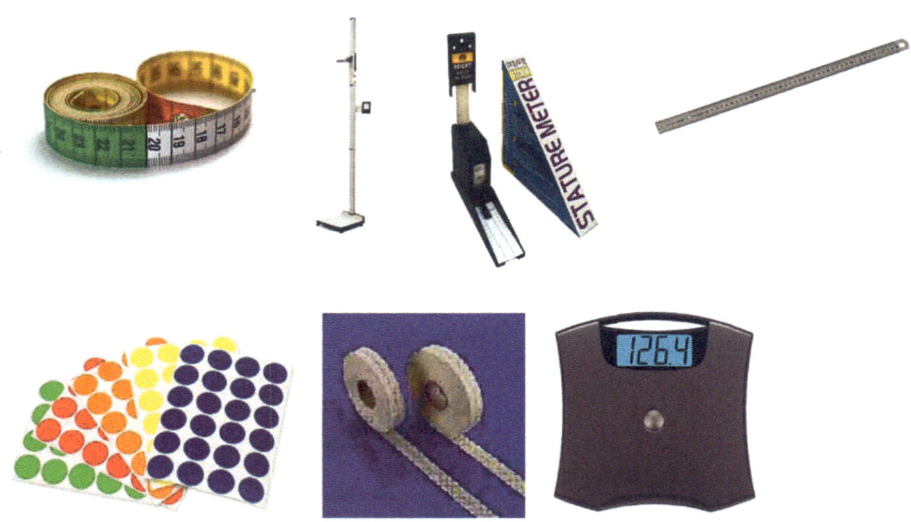

Figure 3.3 Manual anthropometric tools.

Measurement tools used for manual anthropometric survey for the clothing industry were:

- Calibrated nonstretchable plastic measuring tapes (4 pieces)
- Portable and stature meter handy height scale with movable head piece (2 pieces),
- Long ruler (2 pieces)
- Elastic tapes (5 m),
- Round colored circle stickers
- Digital weight scale (2 piece)

The elastic tapes are used to mark the waist and hip for landmark purposes. The circle stickers are used as the landmark on the body whenever necessary. These items can be seen in Fig. 3.3.

For the manual method, the duration to complete one subject is longer than for the digital method as the measurers need to measure 52 body dimensions according to the ISO 8559:1989. Each measurement process takes approximately 30–40 min depending on how well the subjects cooperate at the time they are measured. Furthermore, in order for the measurement process to be conducted smoothly, the cooperation and commitment from the school administration is very crucial.

The school administrators need to be able to give a space in the school for usage during the anthropometric survey. If possible, the team needs to have a quiet place where there is not much disturbance from school activities so that children who are being measured are not distracted by other activities. In addition, the other children who are not involved should not be distracted by the anthropometric survey. The room

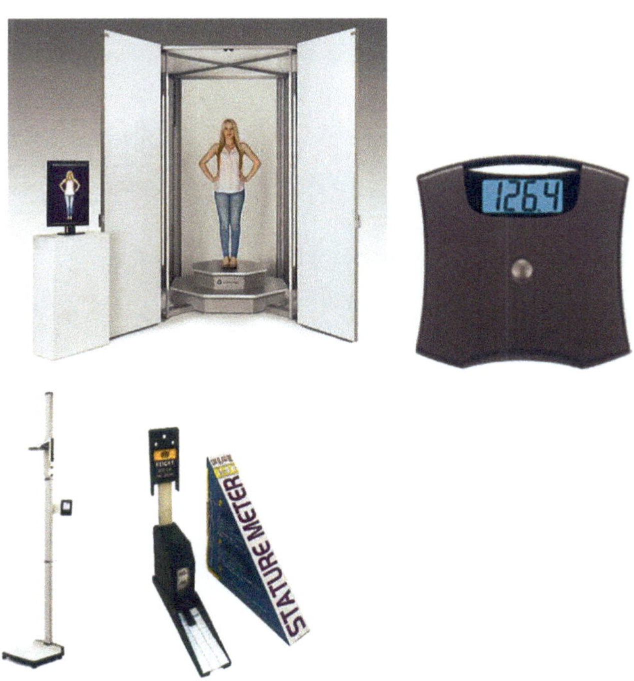

Figure 3.4 Computerized measurement system.

will be a place where the students will sit and wait for their turn to be measured. It should have ample space to put all the equipment, refreshments, and for at least 10 students at one time for measurement. During the measurements, refreshments (simple snacks, drinks, and sweets) are given to the subjects while waiting. The measurement protocols are described next.

3.3.2 Anthropometric manual measurement protocols

The measurement session for each day is suggested to begin at about 9 am, so that the process can be completed by the end of the day at about 5 pm. First, the team should arrive at about 8 am to set up the place, and by 9 am the procedure will start with the selection of subjects from random classes. The measurement process can then start at 10 am. The measurement stops at lunch time and resumes again in the afternoon session at about 2 pm and ends at 5 pm. Measurers are given a break of 10 min between the measurement sessions. Only one person is measured at any one time. The rest of the children wait in a room, while having some drinks and light meals.

The selection of children at the beginning of each session is based on the simple random sampling procedure. For each age group, the student names are obtained from selected classroom. A simple procedure is conducted by the measurers where they write each student's name on a strip of paper. Then, all the strips of paper are put into the basket and shaken. The names are randomly selected from the bowl based on the numbers needed for each age for that day. Finally, the selected names are called in groups of 10 (5 males and 5 females) in order to minimize disruption of classes. The children are directed to a room set aside for the purpose of this research, where they are briefed on the objectives of the survey and asked to change into a tight-fitting garment. A consistent set of procedures is employed for the measurement process, as follows:

- Fill out demographic data such as name, age, gender, and ethnic group on anthropometric survey form
- Measure height
- Measure weight
- Interval: 5 min for children age 7–12 years old
- Measure upper body
- Measure lower body

One measurer will do the measurement and the partner records the measurements. All measurements are taken from one side of the subject body consistently. Afterward, students return to their respective classes and another 10 students are called. This process is repeated until the sample size planned for that day is completed. During the fieldwork for the real anthropometric survey, an average of 30 children can be measured per day for the manual anthropometric survey.

After completing the measurements of the day, the forms are counted and the overall quality of anthropometric measurements is checked. Forms with missing data are considered not valid and are scrapped from the collection. The number of children measured is recorded. This ensures that the targeted number of samples is achieved and that results are accurate.

3.4 Computerized anthropometric survey for children and teenagers

Today anthropometric surveys can also be conducted using technology. This method is also known as the computerized anthropometric survey. The complete system for digitalization of the human body widely popular in the market now is known as the 3D body scanning. The scanning technology was introduced to the clothing industry and became really popular starting from the early 2000s. According to D'Apuzzo, the digital application of technology for measuring the human body is fast, accurate, and very easy to use [12]. The time taken to measure a large number of people is reduced tremendously. This technology used to conduct an anthropometric survey is also non-invasive and protects the privacy of individuals. Children do not have to be restless as the scanning process itself is very short for an average of only 7–12 s. This method prevents children and teenagers from becoming restless and bored. The only time needed is the time for them to change into appropriate scan wear.

3.4.1 Digital measurement system

Digital measurement systems used for computerized anthropometric survey include:

- 3D body scanner
- Digital weight scale (2 piece)
- Computer linked to the scanner for automatic digital data extraction
- Portable and stature meter handy height scale with movable headpiece (2 pieces),

Computerized anthropometric survey has gone through numerous changes since it was introduced to the clothing industry for the purpose of collecting digital anthropometric data. There are several different digital measurement techniques for the human body:

- laser scanning
- projection of white light patterns
- combination modeling and image processing
- digital manual measurement
- technologies based on other active sensors like the optical lenses for body surface capture

As can be seen, there are now many types of 3D body scanners in the market from the cheapest and the simplest to the more sophisticated ones. The most advanced 3D body scanning technology currently for the clothing industry has the auto-extraction measurements applying the ISO 8559:1989 specifically. The body scanner is able to adopt the ISO 8559 standard into their measurement procedures seamlessly without the intervention of the operator. The advantage of this scanner is that it can detect the landmark according to the clothing industry standard and the body measurements are automatically extracted. On the other hand, the operators can also control the measuring points, known as active markers, on the subjects. This is the only specialization of the body scanner that can be set in the scanner automatically without the intervention of the operator. This is compared to the other systems that use passive markers, which include intervention from operators. With this new specification, less time is taken to extract data for analysis.

3.4.2 Anthropometric computerized measurement protocols

The first step before conducting an anthropometric survey using a 3D body scanner is applying ethical clearance from an ethical board committee or an institutional review board. This step is crucial as there are issues when using a computerized scanning method leading to digital image storage of the subjects. Some of the issues that need to be addressed are the nature of the digital image, the participant's confidentiality, how the digital image will be stored with all other personal data, and lastly how the data will be used after the analysis and how can it benefit the research overall.

Under this protocol, a medical doctor is appointed as the consultant to ensure that children under going scanning procedures will not have any physical and/or psychological effects.

In addition to the approval from the ethical committee board, a comprehensive informed consent form also shall be prepared. This form will detail the purpose of this research, the steps needed to complete the research, and the expected outcomes of the research. The form will be written using laymen's terms to allow the public to understand the research. The form will also ensure children's confidentiality, and parents of the subjects can request to withdraw from the research at any time if they see any impacts on their children. The research is also not for commercial purposes. In this form, there will be a protection act stating that the scanning procedure will not harm the subjects physically. As for privacy concerns, the digital image of the subject will be used only for the purpose of research. Privacy protection will include the protection against facial exposure of each subject by using software imaging technology that can mask specific body parts for privacy protection. This means the subject will remain anonymous. Only extracted data points that will not reveal the facial or identity of the subjects scanned will be used for the research.

The first procedure of scanning will be the registration of the participants. Subjects will register at a first booth where one administrator will be. After registration, each subject will be given a barcode number, which will be used for all the processes in the scanning room. The unique barcode number assigned to each person should be input in every activity such as the registration, when filling out the web anthropometric form, the body scanning process, and when answering questionnaires. One measurer will be assigned to take the height and weight measurements and input it into the web anthropometric system in the computer. After this process is completed, one person will then direct each subject to the area for viewing the entire scanning procedure on video. They must also make sure that the subjects are comfortable.

One person in charge here must also answer any possible questions concerning the scanning process. After registration and video viewing, the subject will move into the changing room chamber to change into the tight-fitting scan wear. Then, the subject is guided into the 3D body scanning booth where they will sit and the measurers will adjust the position of the subject according to the standard. The subject should hold onto the hand railing in the scan booth and not move for about 10 s for the camera to scan the whole body. After the completion of the scanning, subjects will change back into their clothes and go out through the chamber to the waiting area.

3.5 Manual anthropometric analysis

The final step of this stage is to analyze the data as follows:

3.5.1 Data entry

All the data collected are keyed into the computer using SPSS software version 13.0. The subject's name and school are keyed in the row, which is known as the case. The variables are keyed in the columns. The first four columns contain demographic information: gender, ethnic group, age, and geographical school area (urban and rural). These are the categorical data. The other 50 columns contain numeric body measurements in numbers, which are known as continuous data.

3.5.2 Data screening

Data screening includes data entry errors, missing data, or outliers. The entire data set is filtered to ensure that there are no errors or missing data. About 10 cases are determined to be outliers while five had missing data and thus were omitted. Some data entry errors are due to human error in keying in the data; these errors are rectified after comparing against the raw data.

The normality test of the anthropometric data is screened using both graphical and numerical methods. The graphical method looks at the histogram of the distribution of selected data. The numerical method looks at the value of mean, median, skewness, and kurtosis.

The histogram also shows a useful graphical representation of the data. Data is normal if the histogram shows normal Gaussian distribution. It involves data examination referring to the bell shape of the data distribution. Selected key variables, namely the height, weight, chest, waist, and hip girths, are illustrated using the histogram with a normal probability curve. This curve should be bell shaped to show normal distribution. Frequencies and normal curve of selected variables are illustrated for male and female populations (7–17 years old). These selected variables are commonly used in garment production.

Using the numerical method, the mean and median value should have the same value while the skewness and kurtosis should show values of 0 and 3, respectively, to indicate that the data is normal [9]. Skewness refers to the asymmetry of the distribution. If the skew has a negative value, this means the data showed skewness to the left; if positive, the skew shows the tail to the right. Kurtosis refers to the peakness or the flatness of the graph.

3.5.3 Data group

The statistical method applied in this step is descriptive analysis, namely the univariate analysis according to simple statistics. Categorical and continuous data are analyzed as described in the following sections. As a whole, the survey collected two types of data namely the categorical (demographic data) and continuous data. After the completion of the whole survey, the data is processed according to the following: data entry and data screening.

3.5.3.1 Categorical data

The categorical data are first analyzed to show the demographic profiles of the samples. The entire data set is divided into two subsets, male and female. Only frequency distribution is described in this section. The separate data are presented according to frequency distribution ordered by quantity and percentage. Data are illustrated using tables and bar graphs.

3.5.3.2 Continuous data

Continuous data are analyzed using simple statistical methods and summarized in the form of tables and histograms. These descriptive statistics include frequency distributions, range, mean, median, mode, standard deviation, coefficient of variation, and Pearson correlation coefficients to determine the interrelationships between the various body dimensions.

Consequently, the first objective of the anthropometric analysis is to profile the demographic data from the categorical data. From the continuous data, the overall pattern of body dimensions is described and distinguished between genders and different age groups. The description of data analysis from body sizes and shapes classification to the sizing analysis is discussed in Chapters 4–9. As a whole, the survey collected two types of data namely the categorical (demographic data) and continuous data and this is described in Chapter 4 onward.

3.6 Computerized anthropometric analysis

The scanning process cuts short the data entry as the measurements are obtained digitally. The body data is extracted from the system to the particular software that can morph the body shapes. The body data can also be extracted to be converted into computerized human body modeling where the different shapes can easily be tabulated visually. This will enhance the findings of the different body sizes, shapes, and proportions of the children as can be seen in Fig. 3.5.

For data analysis, the methods below are proposed:

- Basic descriptive statistical method: average, median, mode, extreme values, and standard deviation (SD). Tabulating the descriptive statistics like average, extreme values, and SD, will give an overview of the shapes and sizes of children.
- Basic statistics such as ANOVA and t-test method will be applied. This statistical method is used to identify any age differences between anthropometric measurements in females and males and within the whole sample. The t-test will be use to evaluate the differences between females and males according to age.
- Data mining method: multivariate, clustering, and classification technique for size classification and determination of different body types. Using the multivariate PCA method, the key

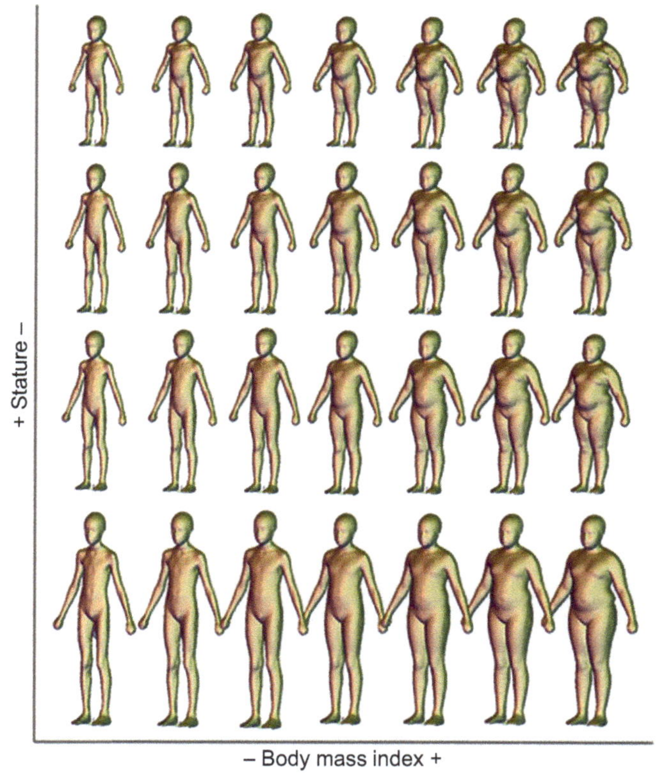

Figure 3.5 Parametric model of child body shapes in standing postures. Adapted from: https://www.google.com/search?q=Parametric+model+of+child+body+shapes+in+standing+postures&espv=2&biw=1745&bih=837&source=lnms&tbm=isch&sa=X&ved=0ahUKEwjrg-b6GkPnLAhWGuo4KHUThAIoQ_AUIBigB#imgrc=gJ-kD1gEugHHlM%3A.

dimensions that reflect the body dimensions of the sample size are identified. From there, the clustering technique will further segment the sample size population into their own body dimensions groups, resulting in distinct body shapes and sizes based on key dimensions. Finally, a classification technique will be used to classify the groups of the body shapes into the right sizes based on the size range, gender, and region.

- Artificial intelligence methods will be used to randomize the data into their own specific size groups in which the accommodation rate of each size group can be determined to ensure the most optimized sizing system.

3.7 Conclusions and recommendations

Conducting an anthropometric survey for children and teenagers is very challenging as the subjects are quite young to understand the seriousness of the research. Careful planning and very systematic procedures need to be considered. In addition, most important is to select team members who can work really well around children and teenagers. In this chapter, the procedures to collect the anthropometric data are described, as well as the actual process of the survey, which is the core fieldwork application. Many different aspects should be taken seriously so as to ensure the smoothness of the fieldwork and also to ensure success of building and collecting the children's anthropometric data for analysis. The aim of the survey is to contribute to a very reliable and accurate sizing system for children's wear and teenager's wear. Some of the main aspects that need to be considered when conducting anthropometric research on children and teenagers are listed herein:

- A short presentation of the measurement objectives needs to be given to children and teenagers before the measurement process. The nature of children, who are always inquisitive and curious, needs to be addressed by giving this presentation. This is to ensure minimal obstruction during the measurement process. If subjects ask a lot of questions during the process, this will delay the overall measurement process. It can also distract from the concentration of the measurers who need to measure accurately. This is even more crucial if manual anthropometric data are to be collected.
- Team members need to be very friendly and firm when dealing with children.
- Impatient young children need to be treated with patience and compassion, which comes from measurers having some knowledge about how to interest children and young teenagers in the process at hand.
- Children can move a lot during the manual measurement process. For some age groups like those 7–9 years old, measurements need to be taken in two stages. They will be given a break of 5 min before continuing again. There is a need to remind them gently to stand still and make them understand that the process will end sooner if they stand straight without moving.
- Measurers for children need to be very communicative with children as the process could be long and monotonous, which might occur when doing manual measurements.
- Teenagers are sometimes more difficult to handle during manual measurements as they may be very shy about being measured as they can have different body issues. For example, they might be very ticklish when being touched and measured.
- Scanning procedures can have different impacts on different children. Children and their parents need to receive constant reminders that the equipment is safe.
- Parents may want to try the scanning procedures to be convinced that it is safe for their children.

References

[1] V. Menon, Size matters, just not in clothing, in thestar.com 19th March, (2009) p.10.
[2] O. Abdali, P. Eric, R. Marc, Exploring anthropometric data through cluster analysis, Int. J. Consum. Stud. 32 (2004) 276–284.
[3] R.S. Bridger, Introduction to Ergonomics, CRC, 2003.
[4] P.M. Jones, M. Rioux, Three-dimensional surface anthropometry: applications to human body variations, Opt. Lasers Eng. 28 (2) (1997) 89–117.
[5] Z.B. Azouz, M. Rioux, C. Shu, R. Lepage, Characterizing human shape variation using 3D anthropometric data, Vis. Comput. 22 (2006) 302–314.
[6] S. Pheasant, Body Space, Anthropometry Ergonomics and Design: Principles Practice of Anthropometry, 1986, pp. 17–18 (Chapter 2).

[7] C. Ruff, Variation in human body size and shape, Annu. Rev. Anthropol. 31 (2002) 211–232.
[8] J.A. Roebuck, Anthropometric Methods: Designing to Fit the Human Body, Human Factors and Ergonomics Society, Santa Monica, CA, 1995.
[9] J.F. Hair, R.E. Anderson, R.L. Tatham, W.C. Black, Multivariate Data Analysis, fourth ed., 1998.
[10] J.E. Bartlett, J.W. Kotrlik, C. Higgins, Organizational research: determining appropriate sample size in survey research, Inf. Technol. Learn. Perform. J. 19 (2001) 43–50.
[11] R.L. Scheaffer, W. Mendenhall, L.R. Ott, Elementary Survey Sampling, 2005.
[12] N. D'Apuzzo, 3D body scanning technology for fashion and apparel industry, in: J. Beraldin, F. Remondino, M.R. Shortis (Eds.), Videometrics IX, Proceeding of SPIE-IS&T Electronic Imaging, vol. 6491, SPIE, San Jose (CA), USA, 2007.

Children and teenagers body sizes and shapes analyses

4

4.1 Introduction

This chapter presents the Sizing analysis stage. The first step is the analysis of anthropometric data obtained from the anthropometric survey that was conducted and discussed in Chapter 3. The process flowchart is shown in Fig. 4.1. As can be seen from this figure, the process of analyzing the anthropometric data involved two methods. One is to analyze the categorical data, which will result in demographic descriptive. The other one is to analyze the continuous data, which gives the body size and shape descriptive.

The objective of this analysis is to achieve the demographic profile and the body sizes and shapes profile of the sample population. The methodology of this chapter is to use simple statistical analysis to examine the data obtained. The goal of this chapter is to show how the anthropometric data is processed and what analysis is being conducted in order to get the output.

In this chapter, Section 4.2 presents the body shape and size in children and teenagers. Section 4.3 presents the body shape and size for children and teenagers. In Section 4.4,

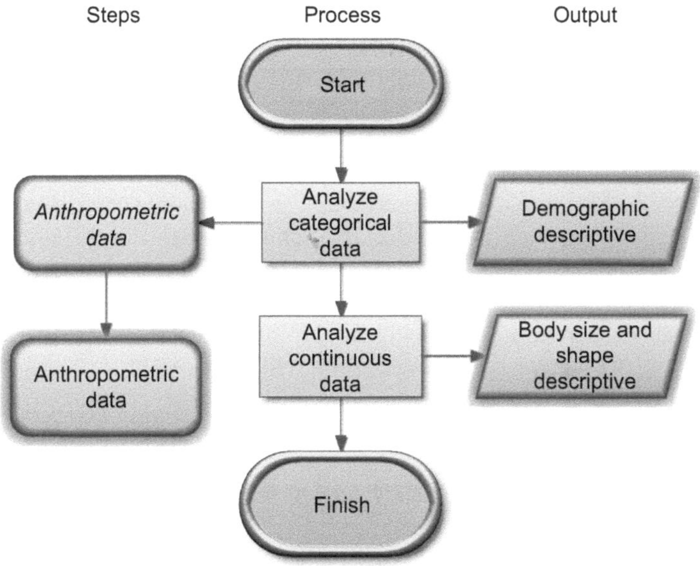

Figure 4.1 The process for anthropometric data.

Clothing for Children and Teenagers. http://dx.doi.org/10.1016/B978-0-08-100226-1.00004-7
Copyright © 2016 Elsevier Ltd. All rights reserved.

the categorical data results for the sample population are compiled. In Section 4.5, the continuous data results are presented according to gender. In Section 4.6 the continuous data results are presented according to gender and age. Lastly, in Section 4.8, the conclusion is presented.

4.2 Body shape and size in children and teenagers

In general, the anthropometric data obtained for this study to understand the general overview of the variations of body sizes, body variations, and body proportions of the children ages 7–17 in one country. The body dimensions can vary in terms of sizes, shapes, and proportions [1]. So far, the anthropometric data have been extensively collected and studied for many different purposes like ergonomics, medical, health, designs, and nutrition and also for the development of sizing systems for apparel manufacturing. These data provide a wealth of information about body dimensions, which is useful for the development of sizing systems for clothing [2].

As early as 1827, studies found that form and variation are the two components by which to classify the human body [3,4]. Later, in 1932, Huxley's [5] research on body variations showed that numerical body dimensions can describe body shapes. In addition, Lele and Richster [6] also mentioned that the relationship between the important body dimensions such as hip, bust, and waist can be interpreted as indicators of shape differences. Many years later, another study reaffirmed the fact that the human body can be classified in terms of body shapes [7]. Interestingly, body shapes have been investigated for many reasons, including health, physiology, understanding of physical aspects, perceptions of attractiveness, and certainly body image and clothing fit [8,9].

The significance of body shape is also beginning to be recognized in the apparel industry for the development of sizing systems [10–12]. This is supported by Gupta [13] who stated that body shape analysis is necessary for the development of accurate apparel sizing. In addition, several studies have revealed that sizing systems must be based on body size and shape and not on age to provide a good fit [14–18]. Hence, manufacturers must ensure they meet the demand of consumers by delivering clothes that have a good fit, are the right size for different body sizes and shapes, are pleasing to the eye, and catch the attention of consumers [19].

Some of these issues are related to body self-perception of children and adolescents; as discussed in the following section, it has been proven that if children are dissatisfied with their body shape, they will likely have problems with their clothing fit [20]. There is also evidence that most obese or plus-sized children are not satisfied with their body shapes and often find it difficult to find the right-sized clothing [21].

Many studies suggest that the establishment of preferred body shapes begins in early childhood [7,22–24]. Previous research discovered that not only are adults conscious of their body image and shape but also children are [20,25,26]. Findings show that adolescents have a fear of being overweight and experience body shape dissatisfaction [18,27]. In another study, results revealed that children as young as age 5 show a desire to be thin with a tubular body shape and perceive thinness as the "right" body shape. Sorensen et al. [28] asked young children their preferences of the ideal body shape, and they chose the underweight figures. When the same question was posed to teenagers, their preference changed from underweight to the average weight figures.

There is also evidence that body image is of special concern for adolescents, not only in females but also in males [29,30]. Chen et al. [20] claim that many adolescents desire to be thin but that there are distinct differences between genders. The findings are consistent with other studies conducted in both Western and Eastern cultures, in which girls show more concern about their body shape and are more dissatisfied with their body shape [31]. While girls showed an inclination toward wanting to be thin, boys showed the opposite desire, associating being larger with being muscular and having physical prowess [32,33]. Girls generally want to be smaller. Boys may want to be thin, but they also want to increase muscle mass [28]. Just as females see an ideal figure in the hourglass shape, males in the United States want to achieve a trapezoid shape [33].

It is essential to understand that previous research has shown that at a very early age, children are already sensitive toward the ideal body shape [34]. This means that they want to wear clothes that can enhance their body shapes as proven in the study by LaBat and De Long [35], who found that when clothing does not fit, the consumer may perceive the cause as related to their body and not the clothing. Another study highlighted that adolescents as early as 9–15 years old go through physical changes that occur in different phases and at different rates. These physical changes affect the fit of clothes [36].

However, Salusso [37] suggested that rather than concluding people's bodies are disproportioned, it is time to seek validity in sizing proportions. Many others have reported that the rapid growth of adolescents not only results in sudden physical changes but also affects their self-perception and social psychology [38]. Social psychology and physical appearance in children are very much related; there is evidence from the literature that children may feel dissatisfied with their body image, which is actually a mental picture of their body rather than their real body shape. Therefore, the ultimate appearance and the fit of clothing must not only be comfortable but must also meet the expectation of body image that pleases the wearer [36]. In addition, with body image strongly connected to body size and shape, it is important that the sizing is based on actual body size and shape, not on age, in order for a garment to fit well [39].

The body sizes showed that there is a difference of height and weight within different genders. From ages 11–12 toward the adolescent years (puberty), the female mean height is higher than the male mean height. Females enter puberty between ages 8 to 13 [40,41]. This is In contrast, males start to have their major growth spurt from ages 10 to 15 years old [42,43]. This phenomenon is associated with puberty ending and sexual maturity in males, which eventually slows down as height is relatively the same from 15 to 17 [44,45]. This is probably due to the fact that the female children reached puberty earlier than age 15 and therefore have stopped growing steadily in terms of height [46,47]. Boys continue to grow until age 16 or 17 as their puberty age stops around those ages. Thus, there is a clear trend from the findings of the research that there is steady growth in height for male samples all the way from age 7 until 17. As compared to females, the steady growth in height occurs from age 7 to 15 and then there is a very slow growth from age 15 to 17 as age increases. The finding also shows that female's growth of horizontal body dimensions from age 7 to 12 is rapid and steady.

4.3 Anthropometric data analysis for children

This section describes the process of anthropometric data, which is divided into two types of data. Fig. 4.1 shows that this study has collected two types of data: categorical and continuous variables. The data were analyzed according to type and gave different statistical descriptive information as shown in Fig. 4.1.

The categorical variables represent some information about the population characteristics of one country such as the age, gender, geographical area, and ethnic group, which can be seen from Table 4.1. For categorical data, the analysis defines the demographic of the sample population. Thus, this analysis would describe the children sample population, which includes the distribution based on total population.

On the other hand, the continuous variables represent some scale, which are the measurements of the body dimensions as described in Table 3.4. Practically, there are 50 different body dimensions taken on each sample. In this chapter, the first basic quantitative analysis conducted is the univariate analysis, which is used to discover the profile of body dimensions of the sample population. In addition, this chapter also

Table 4.1 Geographical school areas and ethnic group distribution

Stratum group	**Geographical school areas**		**Ethnic groups**		
Age/years	**Urban**	**Rural**	**Race A**	**Race B**	**Race C**
7	119	106	76	77	72
8	133	85	73	76	69
9	114	96	76	68	66
10	152	67	70	71	78
11	142	72	73	71	70
12	147	66	68	73	72
13	90	74	54	56	54
14	88	75	56	56	51
15	87	67	50	54	50
16	70	65	45	45	45
17	70	50	47	44	29
Total	**1212**	**823**	**688**	**691**	**656**
Gender					
Male	624	410	350	356	328
Female	588	413	338	335	328
Total	**1212**	**823**	**688**	**691**	**656**
Percentage (%)	**60**	**40**	**34**	**34**	**32**

conducted the bivariate data analysis and the growth rate analysis of some key body dimensions for all sample populations. At the end of this chapter, the continuous data analysis will finally provide the understanding of body sizes and shapes of the sample population that is relevant to the development of sizing system for better-fit clothing.

4.4 Categorical data results and analysis

In this section, the details of the demographic data collected during the anthropometric survey are presented based on four main variables: age, gender, ethnic group, and school geographical area. In total, 2050 subjects were measured. The data were screened for missing values and errors. After data screening and cleaning, the total number of samples usable for analysis was 2035. The data were keyed into the computer using SPSS software version 13.0. Section 4.4.1 discusses the demographic data according to age and gender distributions. This is the main strata group for the survey. Section 4.3.2 discusses the demographic data according to ethnic group and school geographical area.

4.4.1 Sample population according to age and gender

The number of the sample population measured in the anthropometric survey is shown in Fig. 4.2 according to age and gender. The survey has successfully acquired the

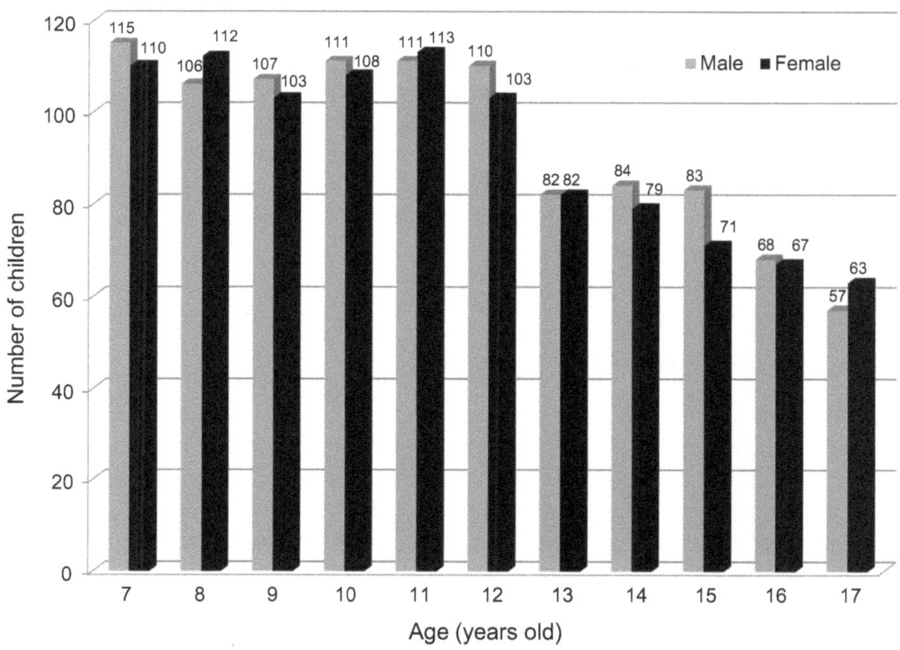

Figure 4.2 Age distribution according to gender.

targeted percentage. Overall, based on the country's total population of children, it showed that the highest number of population of children comes from age 7 and the lowest number of population of children are from age 17. Therefore, the same ratio was taken for the sample population in this study.

In Fig. 4.2, from 11 age groups, eight age ranges have more male samples than female samples. This can be seen in ages 7, 9, 10, 11, 12, 14, 15, and 16. Age 7 had the largest number of male subjects while the female subjects fall under age 8. As seen from Fig. 4.2, the number of samples for age group 13–17 is lower than that for age group 7–12, with 36% of samples falling in the older age group and 64% in the younger age group. This graph represents the real number of population of school-aged children in the country where a bigger percentage of children belong to the elementary age group. Overall, the proportion of male school-age population is slightly higher (51.1%) than the female population (48.9%). Thus, the same percentage is found in the samples totaling 1034 males and 1001 females (33 more male samples than female samples).

4.4.2 Samples according to ethnic group and geographical area

Table 4.1 shows the distribution of samples according to ethnic group and school setting. In terms of ethnic groups, Table 4.1 shows an almost equal sample distribution (34%, 34%, and 32%). The breakdown shows that 688 are Race A, 691 are Race B, and 656 are Race C. This percentage does not represent the real distribution in the country. The representation of ethnic groups was based on random selection on a voluntarily basis. As for school geographical area, 1212 (59.6%) of the sample population come from 17 urban schools while 823 (40.4%) come from 12 rural schools. Therefore, the makeup of the sample has fewer subjects from rural schools than from urban schools, which represent the real school numbers in the country. There are more schools in urban areas as compared to the numbers of rural schools.

From the stratum of gender, the number of male subjects from urban schools was 624 while the number from rural schools was 410. The number of females measured was 588 from urban schools and another 413 from rural schools. Thus, the percentages of male and female subjects from rural and urban schools are 51% and 49%, respectively.

4.4.3 Demographic data characteristics

To summarize this section, Table 4.2 shows the demographic details of the data collected from the anthropometry survey. For this study, the sampling represents the total population from the biggest state in a country. These sample sizes are calculated using the stratified random sampling where the strata group are the age and gender. All parameters are calculated according to the proportionate percentage for that age and gender group in one state that has the biggest population in the country. Therefore, the sampling was taken from it so as to represent the country in particular. For each age, the quantity of samples is given according to ethnic groups, geographical areas, and genders. The age of the subjects varied from 7 to 17 years old, with the following

Table 4.2 Demographic characteristics of sample population

Strata	Geographical area		Ethnic group			Gender		Frequency	Proportionate percentage (%)
Age	**Urban**	**Rural**	**1**	**2**	**3**	**Male**	**Female**		
7–7.9	119	106	76	77	72	115	110	225	11.1
8–8.9	133	85	73	76	69	106	112	218	10.7
9–9.9	114	96	76	68	66	107	103	210	10.3
10–10.9	152	67	70	71	78	111	108	219	10.8
11–11.9	142	72	73	71	70	111	103	214	10.5
12–12.9	147	66	68	73	72	110	103	213	10.5
13–13.9	90	74	54	56	54	82	82	164	8.1
14–14.9	88	75	56	56	51	84	79	163	8.0
15–15.9	87	67	50	54	50	83	71	154	7.6
16–16.9	70	65	45	45	45	68	67	135	6.6
17–17.9	70	50	47	44	29	57	63	120	5.9
Total	**1212**	**823**	**688**	**691**	**656**	**1034**	**1001**	**2035**	**100**
Percentage (%)	58.6	41.4	32.2	34.0	33.9	51.0	49.0		
Gender									
Male	624	410	350	356	328	1034		1034	51
Female	588	413	338	335	328		1001	1001	49

age group representing the largest percentage of the sample: age 7 (11.1%), age 10 (10.8%), and ages 11 and 12 (10.5%). Furthermore, the percentage breakdown of each age is given in Table 4.1. The lowest percentage of the total population comes from age 17 years old.

The proportion of male school children is slightly higher (51.1%) than female students (48.9%), for a total of 1034 males and 1001 females. In addition, total male subjects measured from urban schools were 624 while from rural schools, 410 subjects were measured. The number of females measured from urban schools was 588, with another 413 from rural schools. In terms of ethnic group, the sample showed a distribution of almost equal number (34%, 32.2%, and 33.9%). The breakdown shows that 688 are from ethnic group 1, 691 are from ethnic group 2, and 655 are from ethnic group 3. Further analysis shows that 58.6% of the sample population is measured from 17 urban schools while another 41.4% are gathered from 12 rural schools. The total sample population measured in urban school was 1212 while 823 children were measured from rural schools. Thus, 60% of the samples were from rural schools and 40% were from urban schools.

4.5 Continuous data results and analysis for the sample population

In this section, the data of the body dimensions, namely the continuous data, are described. There are three phases of data analyses as shown in Fig. 4.3. The first step was to analyze the data to get an overview of the body dimensions for the

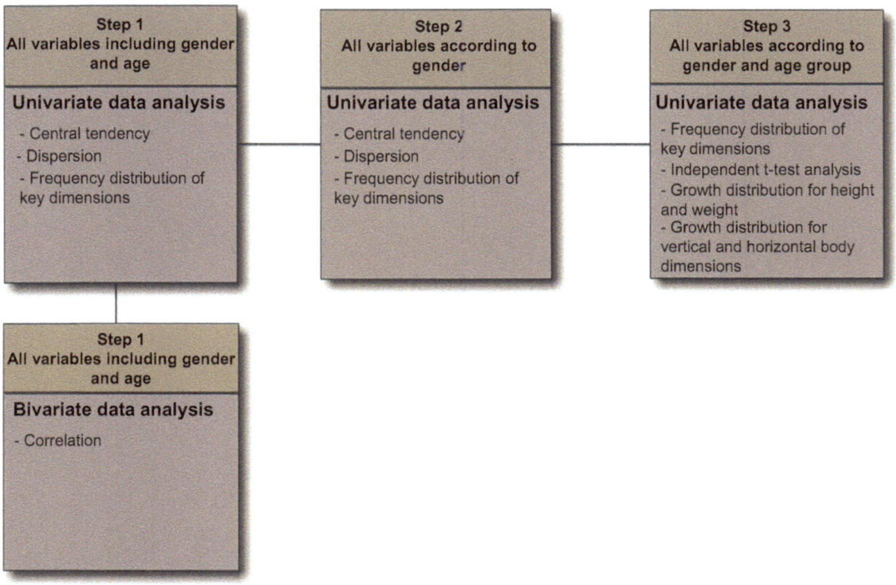

Figure 4.3 Process of univariate and bivariate data analysis.

whole sample population. Here, there were two analyses conducted, which are the univariate and bivariate analyses for all variables using the entire sample population. For Step 2, all variables were analyzed using univariate technique according to gender (male and female subjects). Lastly, in Step 3, all the variables were analyzed using univariate data analyses based on key dimensions according to gender and age.

Thus, Sections 4.5.1–4.5.3 in which the body dimensions data for the entire sample population are examined based on descriptive, frequency, and correlation analyses are based on Step 1. The descriptive body dimensions will provide the profile of the body sizes of the sample population. In addition, the frequency analyses will provide the data distribution of each body dimension and the discussion of the multiple correlation coefficients is to determine the strength and direction of the relationships between the variables.

Then, in Sections 4.5 and 4.6 the anthropometric data is analyzed according to gender which is in Step 2. The descriptive and frequency distribution analyses are conducted based on male and female anthropometric data. Lastly, in Sections 4.6 and 4.7, the anthropometric data is analyzed based on gender (male and female) and age range (children and teenagers) separately using Step 3. The main key dimensions are discussed such as the height and weight distributions of both genders with different age ranges. In addition, the growth of horizontal and vertical body dimensions of both genders and age range are analyzed and discussed in this section. Lastly, Section 4.8 gives the chapter conclusions.

4.5.1 Descriptive data distribution for the sample population

This section describes the profile of body size for the sample population. The descriptive statistics for all the measured body dimensions for the entire sample size ($n = 2035$) are tabulated and shown in Table 4.3.

The dimensions and intervals define the size range of a targeted population in order to provide a reasonable fit. According to Beazley [84], clothing manufacturers always wish to fit as many customers as possible but it is more profitable to produce as few sizes as possible. Therefore, the range of sizes for the key dimensions is determined by calculating the mean, standard deviation, and extremes for the various sample groups.

As can be seen from Table 4.3, the height distribution of 7–17 years old is 101–184.4 cm with a mean of 141.9 cm and a standard deviation of 16.8 cm.

The weight distribution from the same data range from 13.4 to 146.7 kg and the standard deviation is 17.5. Table 4.3 also provides the range of detailed body dimensions for the sample population as knowing the distributions of various body dimensions can help design the clothing sizes that fit them well. If we are designing for clothing we need to understand the body dimensions for sizing distribution for better fit. We also need to understand the range of the body dimensions in a country because each country has different ranges and average sizes. Thus, from the anthropometric survey, it is helpful to first tabulate all of the data based on descriptive analyses. From this table, the understanding of the data can be studied in the

Table 4.3 **Body size characteristics of sample population**

	Sample population (*N* = 2035)				
Body dimensions	**Mean**	**SD**	**CV (%)**	**Min**	**Max**
1. Height	141.9	16.8	11.8	101.0	184.4
2. Weight	40.8	17.5	42.9	13.4	146.7
3. Head girth	53.1	2.4	4.5	34.6	63.0
4. Neck girth	30.0	3.6	12.0	22.5	53.0
5. Neck base girth	34.5	4.0	11.6	25.0	59.0
6. Shoulder length	11.2	1.9	17.0	6.6	19.0
7. Shoulder width	35.1	4.6	13.1	20.0	51.0
8. Back width	27.9	5.4	19.4	14.5	55.0
9. Chest girth	74.0	12.3	16.6	35.3	134.0
10. Upper arm girth	25.0	5.0	20.0	12.6	46.5
11. Armscye girth	35.4	6.1	17.2	20.0	66.5
12. Under arm length	39.0	5.4	13.8	20.4	66.0
13. Upper arm length	28.1	4.0	14.2	18.5	51.0
14. Arm length	51.5	7.1	13.8	32.0	80.5
15. Cervical to wrist	68.8	9.1	13.2	32.0	95.2
16. Elbow girth	23.5	3.8	16.2	15.5	41.8
17. Wrist girth	14.2	1.8	12.7	10.0	27.2
18. Hand girth	17.6	2.0	11.4	12.8	25.5
19. Hand length	16.3	2.0	12.3	11.5	22.1
20. Waist girth	63.8	11.9	18.7	23.5	124.5
21. Waist height	90.9	12.3	13.5	48.5	121.4
22. Outer leg length	91.6	12.4	13.5	53.2	129.1
23. Hip girth	78.2	13.6	17.4	32.5	131.2
24. Waist to hips	18.3	3.4	18.6	9.5	31.0
25. Hip height	73.1	9.8	13.4	37.4	97.8
26. Body rise	25.4	4.2	16.5	13.5	43.0
27. Scye depth	15.6	3.1	19.9	8.7	35.5
28. Cervical to breast point	27.9	4.5	16.1	16.8	47.9
29. Cervical to waist front	40.0	5.4	13.5	24.1	80.5
30. Cervical to waist back	32.1	4.8	15.0	14.5	47.5
31. Trunk length	56.7	8.4	14.8	32.0	83.0
32. Cervical height sitting	54.1	7.7	14.2	36.2	87.3
33. Cervical to knee hollow	83.7	11.6	13.9	42.1	116.5
34. Cervical height	122.8	16.3	13.3	83.1	166.1
35. Neck shoulder to breast point	20.3	3.9	19.2	12.1	34.0
36. Neck shoulder to waist	32.3	4.8	14.9	20.6	48.5
37. Crotch	66.4	12.1	18.2	29.5	123.1
38. Trunk circumference	127.4	19.0	14.9	86.1	197.1
39. Thigh girth	46.9	9.2	19.6	22.5	82.0
40. Thigh length	25.5	3.7	14.5	15.5	50.0
41. Inside leg length/crotch height	64.8	8.7	13.4	30.6	88.8

Table 4.3 **Continued**

	Sample population (N=2035)				
Body dimensions	**Mean**	**SD**	**CV (%)**	**Min**	**Max**
42. Mid-thigh girth	40.3	7.9	19.6	23.8	72.5
43. Knee girth	32.3	5.1	15.8	20.0	54.1
44. Knee height	39.7	5.4	13.6	12.5	58.1
45. Lower knee girth	28.9	4.6	15.9	19.2	48.0
46. Calf girth	29.9	5.2	17.4	17.5	54.5
47. Minimum leg girth	20.0	2.8	14.0	12.8	41.2
48. Ankle girth	23.1	2.7	11.7	0.9	2.6
49. Ankle height	6.2	0.9	14.5	4.0	13.9
50. Foot length	22.9	2.6	11.4	8.8	30.8

extreme values and mean values for sizing expectations. Then range and distribution of the data can be seen from the frequency analysis in the next section.

4.5.2 *Frequency Data Distribution for the Sample Population*

In this section, the data is analyzed using a histogram graph. It provides the frequency distribution for five key dimensions of the body dimensions data: height, weight, chest girth, waist girth, and hip girth. The purpose of using a histogram graph is to understand the distribution of values for each of the key dimensions in order to recognize the pattern of body sizes for the sample population and to see the normal distribution of the data. The five body dimensions selected are the dimensions that are important for sizing development. The frequency distributions will be able to show the actual number of samples falling in each range of body dimensions. For example, Fig. 4.3 showed the numbers of sample population according to different height clusters, which cannot be observed from descriptive analysis. The variations of measurements can also be seen from this analysis as the body dimensions are clustered according to its range.

As can be seen, Fig. 4.4 provides the distribution of height of the whole sample population, which shows a bimodal shape. It is desirable to find the cause of this bimodalism, as height was proven to have a bimodal distribution due to the mixing of separate distinct groups [48]. A possible explanation for this might be that this study consists of female and male samples. There is evidence from many previous studies that a bimodal height histogram may be due to having male and female groups or other factors like different age groups [49,50].

Another study by A'Hearn [51] found that height bimodalism can be caused by many factors such as inadequate nutrition or subjects who are at ages before adult height is reached. As this sample consists of ages 7–17, the height bimodalism in this sample could be due to subjects being at ages where growth is rapidly increasing

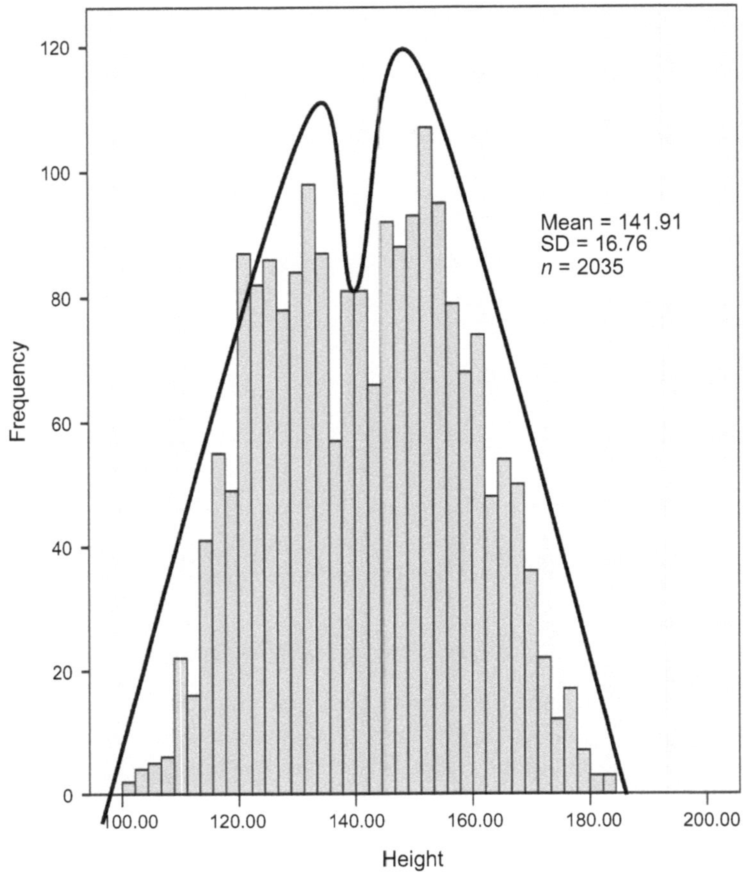

Figure 4.4 Height distribution for males and females ages 7–17.

before final adult height is reached. It is also acknowledged by other studies that the phenomenon of adolescent growth spurt can distort the distribution of normality in height [48].

Thus, with the existence of bimodalism, previous studies suggest that the group should be split into separate subgroups according to the factors in that sample population so that individual analysis can be carried out on each subgroup separately to determine the causative factors [48,49].

From Fig. 4.5, weight ranged from 13.4 to 126.7 kg while the average weight was 40.8 kg. The standard deviation and skewness value are both quite large showing a much skewed distribution, with the samples mostly clustered in the lower weight range.

Fig. 4.6, shows chest girth of the subjects. Chest girth is measurement of the upper body. The smallest chest measures 35.3 cm while the largest chest measures 134.0 cm. Chest girth showed normal distribution and have mean measurements of 74.1 cm.

Key dimensions that are often used for the lower body in a sizing system are waist and hip girth. Fig. 4.7 shows the mean waist measurement is 63.8 cm and the range is

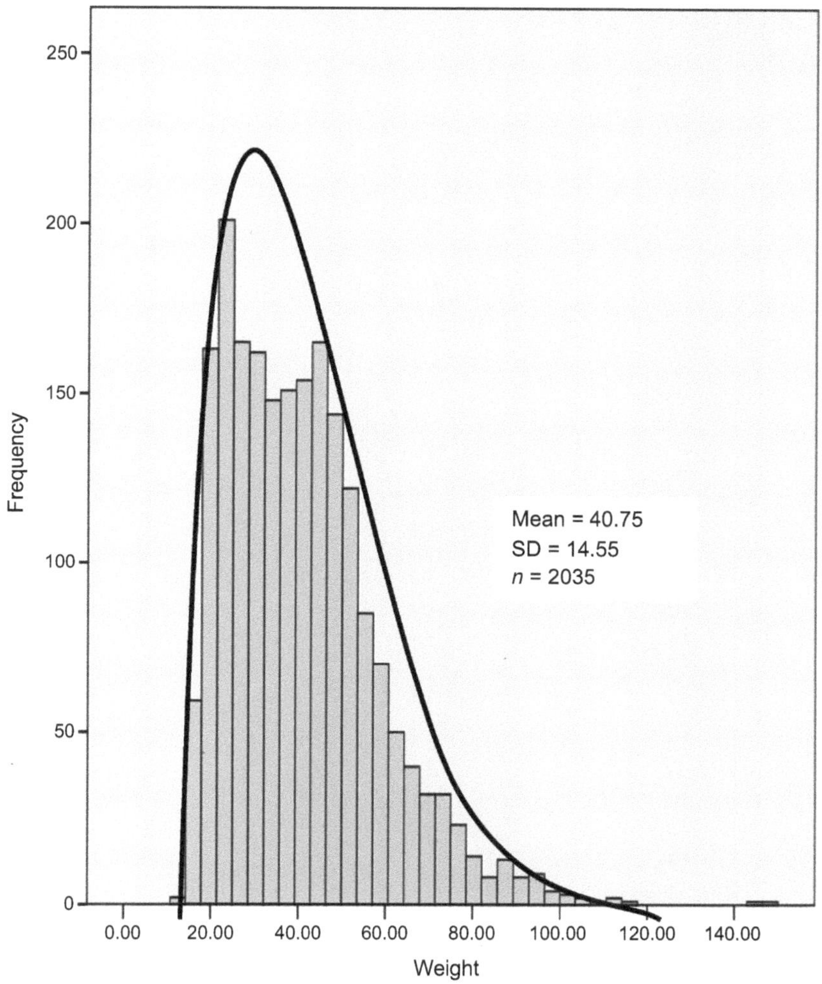

Figure 4.5 Weight distribution for males and females ages 7–17.

113.3 cm. In addition, the smallest waist circumference is 23.5 cm while the largest waist measurement is 124.5 cm. As can be seen from Fig. 4.8, the mean hip measurement for this sample is 78.2 cm with a range of 32.5–131.2 cm. The waist girth distribution is skewed to the left, which means the waist measurements are mostly scattered in the lower range. Hip distribution is the least skewed and shows an approximately normal distribution.

The distribution of key dimensions as shown in this chapter is useful for the estimation of size roll for the sizing system. Overall, the size range for height is 83.3 cm. As for weight, Otieno [52] has mentioned in her study that this dimension is useful only in smaller children. Previous studies clearly suggest that bust and hip girth should be used as key dimensions to replace weight. The reason for this is that both these dimensions (bust and hip girth) are flexible for varying size relationships for upper and lower body [53,54].

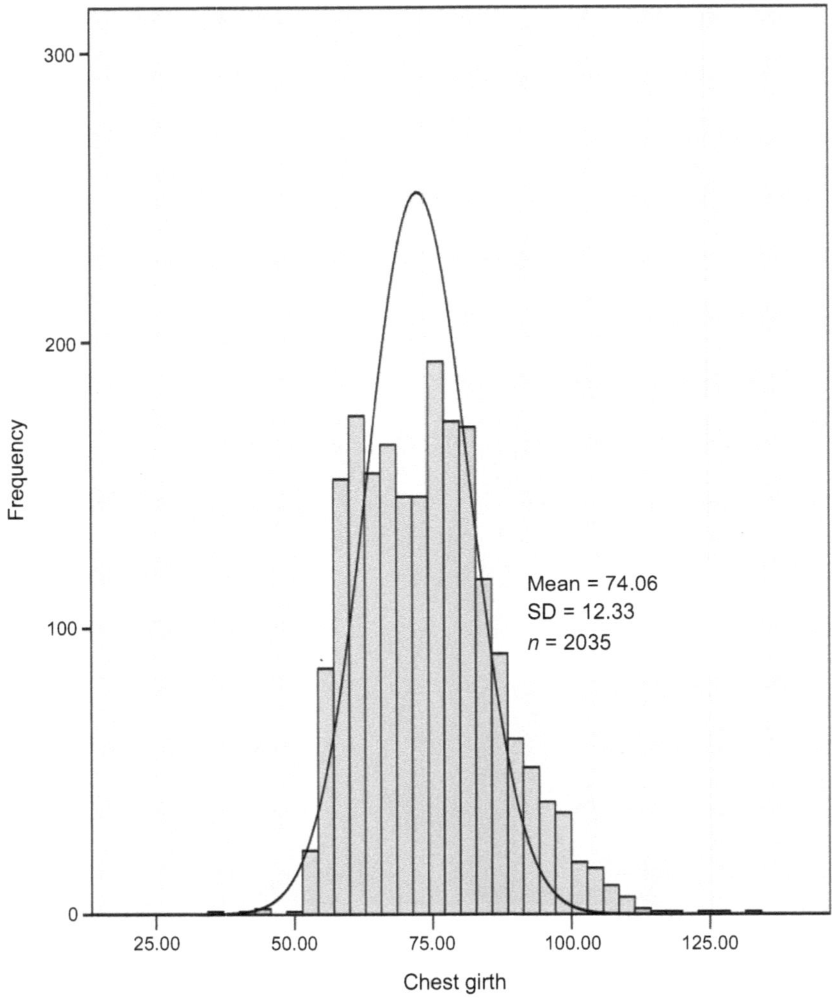

Figure 4.6 Chest distribution for males and females ages 7–17.

Another reason for substitution of weight is that these dimensions (bust and hip girth) are the body dimensions in contact directly with the clothes worn while weight is not [55].

For the upper body, the size ranges for chest and bust girth are 98.7 and 89.0cm, respectively. For the lower body, waist and hip girth measurements indicate size ranges of 101.0 and 98.7cm, respectively. However, these extreme values seem high for the size development. This indicates that the extreme values of the key dimensions may not be suitable for the size range; instead, the percentile value is more suitable. This is supported by previous studies in which the percentile value of key dimensions was used covering 95% of the population [56]. Thus, for this study the percentile values are applied to the key dimensions where the 5th and 95th percentile are used to accommodate 90% of the sample population. This percentile value will be applied to the key dimensions selected in the next stage of this study.

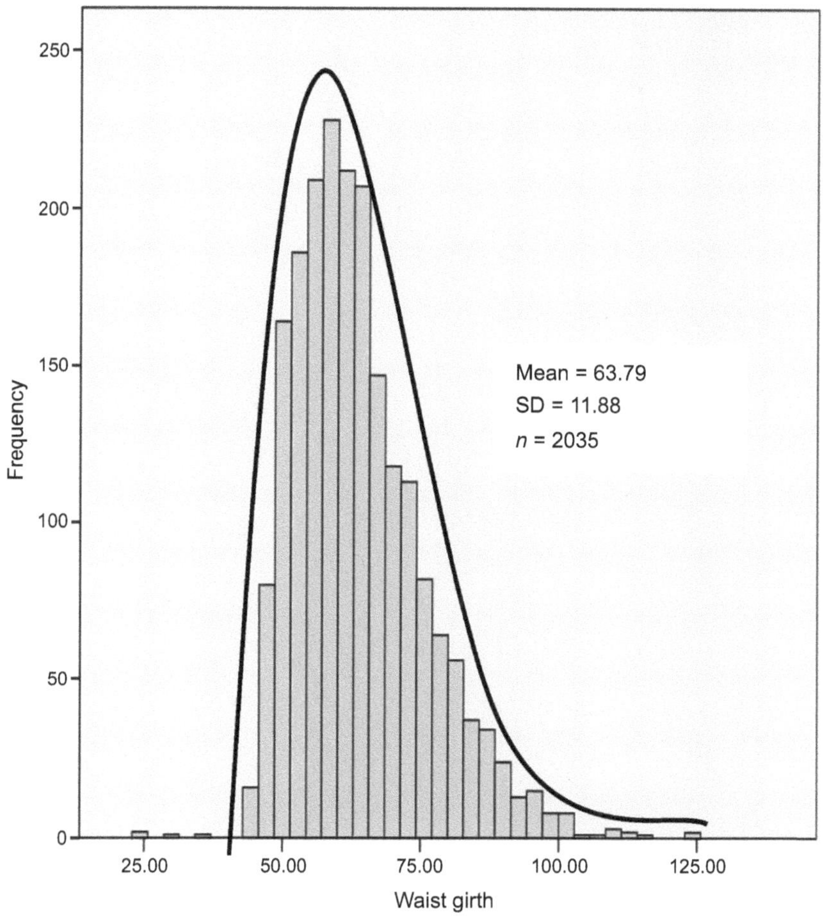

Figure 4.7 Waist distribution for males and females ages 7–17.

4.5.3 Correlation analyses of key dimensions for the sample population

In this section, the correlation analysis technique is used to analyze the data of the whole sample population. The purpose of using Pearson product–moment correlation tests for all body dimensions is to quantify the dependence of two variables. All variables were significant at 0.000 ($n = 2035$). The results indicate that length variables are strongly correlated with other length variables and are girth variables with other girth variables. Thus, increases in length dimensions of the sample population are correlated with increases in other length dimensions and increments in girth dimensions are associated with increases in other girth variables. The finding of this study seems to be consistent with other research that found that height and weight correlated with the most multiple coefficients [54,57,58].

Height and weight, which are commonly used to describe the body size of children, were found to have strong correlations with each other ($r = 0.81$). Height is also found

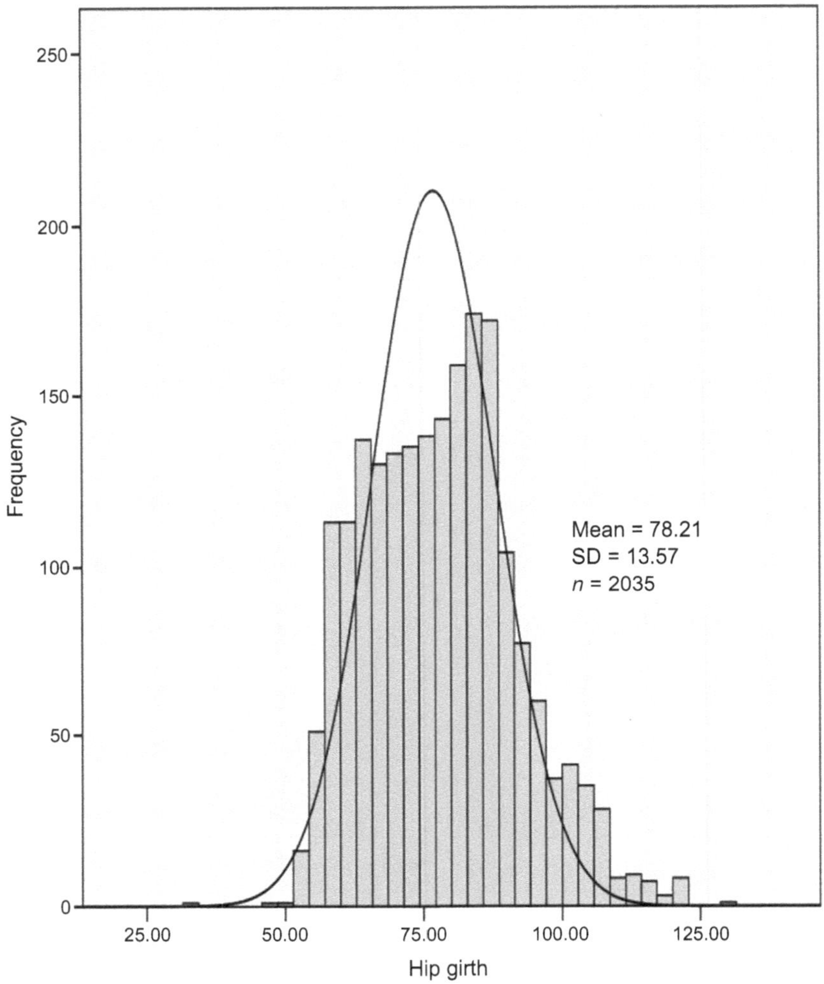

Figure 4.8 Hip distribution for males and females ages 7–17.

to be correlated with all the other 50 dimensions, with 38 dimensions showing a strong correlation and 12 moderately correlated. Height is very strongly correlated with all length dimensions, mainly with the major vertical dimensions like under arm length ($r = 0.93$), cervical height ($r = 0.99$), cervical to waist back ($r = 0.89$), inside leg length ($r = 0.94$), and cervical to wrist ($r = 0.95$), which can be seen in Table 4.4 and Fig. 4.9. This finding is similar to those of other studies where a strong correlation is found between height and length dimensions [57,59,60,63].

Height also has a strong relationship with several girth dimensions, namely chest girth ($r = 0.76$), bust girth ($r = 0.78$), and hip girth ($r = 0.77$). This indicates that as height increases, girth dimensions like chest, bust, and hip girth will likewise increase. However, height was found to be only moderately correlated with waist girth ($r = 0.58$), which shows that waist increment is moderately associated with height increment.

Table 4.4 **Correlation results of important body dimensions**

	Variables	1	2	3	4	5	6	7	8	9	10	11
1	Age	1										
2	Height	0.86	1									
3	Weight	0.67	0.81	1								
4	Chest girth	0.63	0.76	0.95	1							
5	Waist girth	0.43	0.58	0.89	0.90	1						
6	Hip girth	0.66	0.77	0.95	0.94	0.89	1					
7	Under arm length	0.80	0.93	0.71	0.67	0.50	0.68	1				
8	Cervical height	0.85	0.99	0.83	0.79	0.62	0.80	0.93	1			
9	Cervical to waist back	0.80	0.89	0.78	0.72	0.57	0.73	0.81	0.90	1		
10	Inside leg length	0.81	0.94	0.69	0.65	0.47	0.66	0.92	0.94	0.79	1	
11	Cervical to wrist	0.83	0.95	0.84	0.80	0.64	0.81	0.91	0.96	0.85	0.90	1

<0.5 = mild, 0.5–0.75 = moderate, 0.75–0.90 = strong, >0.9 = very strong.

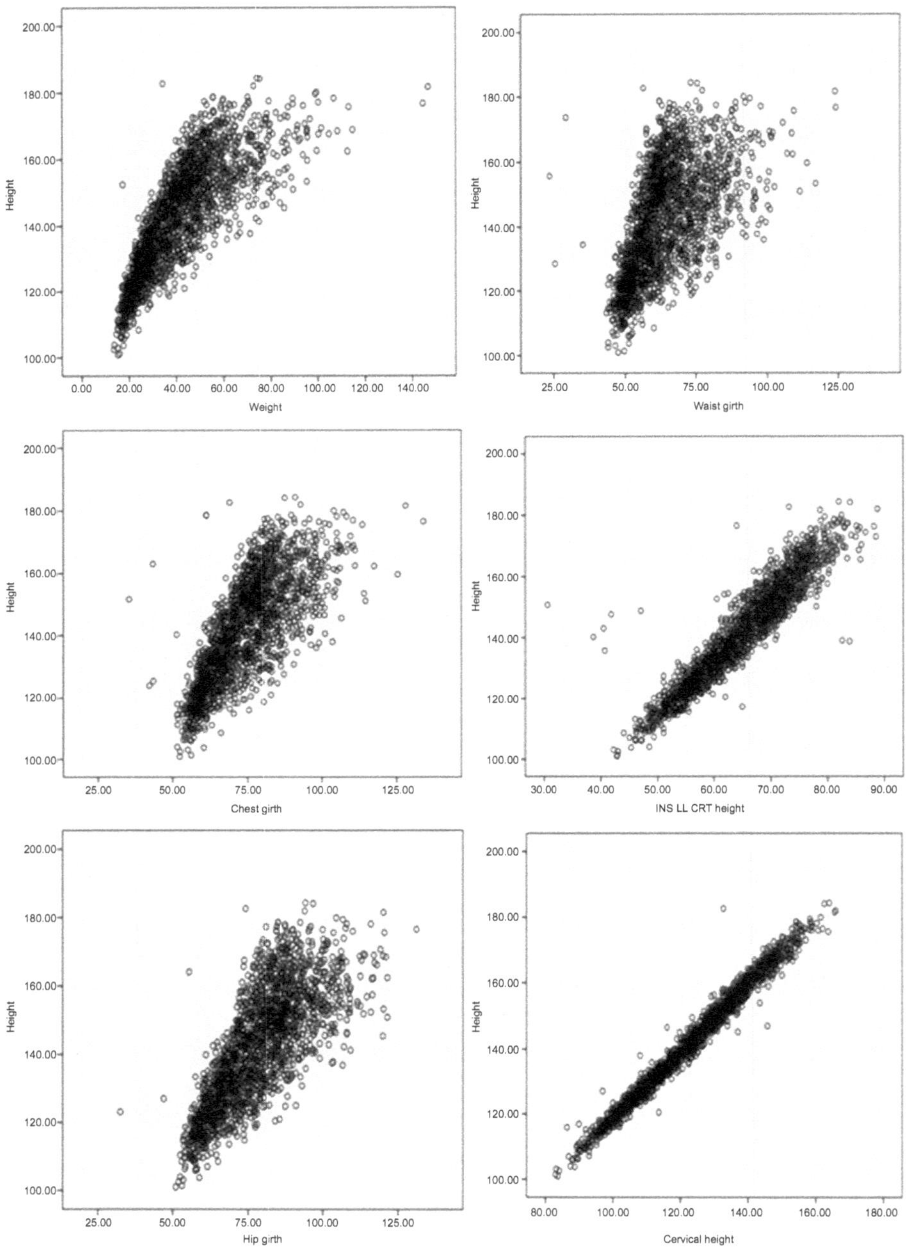

Figure 4.9 Scatter plot for selected variables.

Increases in body dimensions of length and girth are expected as children between ages 7 and 17 go through rapid physical development before reaching adulthood [61]. Height is often used primarily to assess linear growth in children [18]. Moreover, body proportion and composition are also essential elements of growth in children and indicate maturation toward adulthood [14].

These results confirm that height and major girth dimensions (chest, bust, and hip girth) are reliable key dimensions for children ages 7–17. Height, chest girth, and bust girth are often used for upper body classification. In contrast, height, waist, and hip are used for lower body classification [11,15,39,62]. From this study, waist was not found to be highly correlated with height as compared to hip, suggesting that waist may not be suitable for classification of lower body as compared to hip girth.

However, for practicality, not all the dimensions are used for sizing development but instead only the key dimensions are used. As mentioned in the Robinette study [63], key dimensions should be the dimensions that have correlations with most other body dimensions. The stronger the correlation between dimensions, the better suited it is to be a key dimension in classifying the sample population. From this study, weight is found to be moderately correlated with age ($r = 0.76$). However, it is very strongly correlated with other girth dimensions, namely chest girth ($r = 0.95$) and hip girth ($r = 0.95$). Moreover, it is strongly correlated with waist girth. As for length dimensions, weight is highly correlated with height ($r = 0.86$), cervical height ($r = 0.83$), and cervical to wrist ($r = 0.84$). From Table 4.4, it can be seen that weight is also correlated with other dimensions.

It is also observable from Table 4.4 that most of the key girth dimensions like bust, chest, waist, and hip girth were all found to be strongly correlated with each other with r values between 0.90 and 0.98. This result shows the same trend as other studies in which girth dimensions are highly correlated with each other and with weight [14,52,59]. This implies that other girth dimensions, which are more meaningful for apparel manufacturing for children ages 7–17, can adequately represent weight. This is due to the fact that weight is often used for smaller children from babies to toddlers for sizing tables and thus is not suitable for older aged children [77,85]. Thus, this study confirms that height can be the representative for length dimensions since it is correlated with all body dimensions. The girth dimensions that can best represent weight are bust, chest, waist, and hip girth since they are highly related to weight and all other girth variables.

4.6 Continuous data results and analysis for the sample population according to gender

In the next two sections, the anthropometric data set will be analyzed according to the two genders, male (7–17 years old) with $N = 1034$ and female (7–17 years old) with $N = 1001$.

4.6.1 Descriptive data analysis according to different gender

For this section, the first analysis for the anthropometric data is the simple statistical descriptive statistics where mean, standard deviation, correlation coefficient, and extreme values like minimum and maximum were calculated and compared for each variable according to gender, as presented in Table 4.5. Overall, significant gender differences were found in all of the 43 body dimensions marked with

Table 4.5 **Descriptive statistics of anthropometric data according to gender**

	Male ($N=1034$)					Female ($N=1001$)				
Body dimensions	**Mean**	**SD**	**CV**	**Min**	**Max**	**Mean**	**SD**	**CV**	**Min**	**Max**
**Height	143.2	18.2	12.7	101.0	184.4	140.6	15.1	10.7	103.0	166.1
**Weight	42.2	19.1	45.3	14.5	146.7	39.3	15.4	39.3	13.4	91.8
Head girth	53.1	2.5	4.7	34.6	62.0	53.1	2.3	4.4	46.1	60.4
Neck girth	30.1	3.9	13.0	22.5	53.0	29.9	3.1	10.4	23.0	42.0
**Neck base girth	34.3	4.4	12.8	25.0	59.0	34.6	3.5	10.1	26.0	45.0
**Shoulder length	11.3	2.1	18.6	6.6	17.9	11.1	1.8	16.2	7.0	19.0
**Shoulder width	35.4	5.0	14.1	24.0	51.0	34.8	4.2	12.1	20.0	43.5
**Back width	26.1	5.4	20.7	14.5	55.0	29.8	4.6	15.3	19.0	41.5
**Chest girth	74.5	13.1	17.6	35.3	134.0	73.0	12.7	17.4	32.5	108.0
**Upper arm girth	24.7	5.2	21.1	12.6	46.5	25.3	4.7	18.6	15.4	42.0
**Armscye girth	35.7	6.4	17.9	21.5	66.5	35.1	5.9	16.8	20.0	57.0
**Under arm length	39.4	5.8	14.7	20.4	52.0	38.6	5.0	13.0	25.0	66.0
Upper arm length	28.1	4.1	14.6	18.7	39.0	28.2	3.8	13.6	18.5	39.0
**Arm length	52.0	7.6	14.6	32.0	80.5	51.1	6.6	12.9	32.0	65.0
**Cervical to wrist	69.1	9.7	14.0	32.0	95.0	68.4	8.5	12.5	32.5	85.5
**Elbow girth	24.1	4.1	17.0	15.5	41.8	22.8	3.3	14.5	15.5	32.5
**Wrist girth	14.4	1.9	13.2	10.0	23.3	14.0	1.6	11.3	10.4	27.2
**Hand girth	17.8	2.2	12.4	12.8	25.5	17.3	1.7	10.0	13.0	21.0
**Hand length	16.6	2.1	12.7	12.0	22.1	16.1	1.8	11.0	11.5	19.0
**Waist girth	64.7	12.9	19.9	23.5	124.5	62.8	10.7	17.0	44.0	101.0
**Waist height	91.8	13.4	14.6	48.5	121.4	89.9	11.0	12.2	59.8	108.5
**Outer leg length	92.7	13.5	14.6	53.0	129.0	90.5	11.1	12.3	59.5	110.0
**Hip girth	77.1	13.6	17.6	32.5	131.2	79.3	13.4	16.9	47.0	120.0
**Waist to hips	18.9	3.5	18.5	9.5	31.0	17.7	3.2	18.3	10.0	25.0
**Hip height	73.8	10.5	14.2	48.2	97.8	72.6	9.0	12.3	37.4	88.0

**Body rise	26.3	4.6	17.5	13.5	43.0	24.5	3.5	14.1	17.0	32.0
**Scye depth	17.2	3.1	18.0	8.7	35.5	14.0	2.2	15.7	9.0	32.0
Cervical to breast point	27.8	4.5	16.2	16.8	47.0	28.1	4.5	16.1	19.0	39.0
**Cervical to waist front	39.6	5.3	13.4	24.0	80.5	40.4	5.5	13.5	34.5	64.5
**Cervical to waist back	32.6	5.1	15.6	22.0	47.5	31.5	4.4	14.1	14.5	40.5
**Trunk length	58.1	9.1	15.7	32.1	83.2	55.2	7.2	13.1	32.5	70.0
Cervical height sitting	54.4	8.1	14.9	36.3	80.8	53.8	7.3	13.5	36.2	86.6
**Cervical to knee hollow	84.8	12.4	14.6	56.2	116.5	82.7	10.6	12.8	42.0	100.0
**Cervical height	124.4	17.7	14.2	83.2	166.2	121.2	14.7	12.1	83.2	147.5
*Neck shoulder to breast point	19.5	3.8	19.5	12.3	34.2	21.1	3.9	18.5	13.0	30.0
**Neck shoulder to waist	31.2	4.5	14.4	20.6	44.5	33.4	4.9	14.5	23.5	47.0
**Crotch	68.6	13.1	19.1	29.5	123.4	64.1	10.5	16.4	42.0	100.0
**Trunk circumference	128.5	20.3	15.8	87.3	197.2	126.3	17.7	14.0	86.0	181.0
**Thigh girth	46.5	9.6	20.6	22.5	79.1	47.3	8.7	18.5	23.5	72.5
**Thigh length	25.8	4.1	15.9	15.5	50.2	25.3	3.4	13.3	15.5	34.5
**Inside leg length/crotch height	65.4	9.4	14.4	30.6	88.8	64.2	7.9	12.3	40.7	78.1
Mid-thigh girth	40.5	5.3	13.1	23.8	72.5	40.0	7.3	18.4	25.1	63.2
**Knee girth	32.6	5.3	16.3	22.1	54.1	32.0	4.8	15.1	20.2	49.3
**Knee height	40.5	5.9	14.6	12.5	58.2	39.0	4.6	11.8	22.6	47.5
Lower knee girth	29.0	4.9	16.9	19.2	48.3	28.8	4.3	15.1	19.2	43.5
Calf girth	30.0	5.5	18.3	19.3	54.5	29.8	4.9	16.3	17.5	43.5
Minimum leg girth	20.1	3.0	14.9	12.8	41.2	19.9	2.5	12.8	14.1	33.4
**Ankle girth	23.6	3.0	12.7	15.5	35.1	22.6	2.3	10.3	16.3	28.5
**Ankle height	6.4	0.9	14.1	4.1	8.8	5.9	0.8	13.5	4.1	8.2
**Foot length	23.1	2.8	12.1	16.4	30.4	22.6	2.2	9.8	13.9	28.8

** Statistically significant differences between males and females ($p<0.001$). Chest is measured for both male and female.

asterisks (*) in Table 4.5. Generally, this study reveals that most of the males' body dimensions have higher value in body measurements except in the following 10 dimensions: neck base girth, back width, upper arm girth, upper arm length, hip girth, cervical to breast point, cervical to waist front, neck shoulder to breast point, neck shoulder to waist front, and thigh girth. Previous studies found that hip girth was the only dimension smaller in boys and waist was bigger in males, which is consistent with the findings of this study [64].

The findings from this study also show that females have bigger dimensions in the thigh and hip area, which may be accumulated by female higher gynoid or female fat in that area. This supports previous studies that showed that overall girls will accumulate more fat in the breasts, hips, and thighs as the hormones increase during puberty and toward adulthood [61,65].

Table 4.5 also shows that overall male subjects have larger standard deviation in all dimensions, ranging from 0.1 to 3.7 cm, except in cervical to waist front, neck shoulder to breast point, neck shoulder to waist, and mid-thigh girth; these showed higher values in female subjects. In addition, the majority of the female body dimensions have lower values as compared to male subjects. In contrast, the maximum values are all larger in male subjects except for shoulder length, under arm length, wrist girth, and cervical height sitting.

This study indicates that overall, the majority of the body dimensions are bigger in the male sample population based on this country's population. In addition, the lowest value of coefficient variation (CV) is in head girth while the highest CV is in weight for both genders. The CV values for variables were about the same for both genders in comparison to each gender separately.

From the table, it can be seen that out of 50 body dimensions, there are only 21 variables with the high CV. Coefficient variation is used rather than standard deviation (SD) to indicate the variation of the body dimensions. CV is the estimated SD divided by the mean of each variable. The variations are wider when the CV is higher. From this analysis, the result showed that weight has the highest CV, meaning that for both samples, weight has the widest variations. This table also showed that although the SD is high, it does not mean that their CV is high in comparison. For example, height has SD 18.15, but the CV is only 12.15, which is not that high as compared to other variables that have high CV of about 19. For the other dimensions, trunk girths (namely bust, waist, hip, and thigh girths) also showed high CV that have wide variability. Thigh girth showed the highest variation among the four trunk girths but is still relatively smaller than the weight variation.

Ten length measurements showed high variability with crotch distance giving the highest variation. It should be noted that these body dimensions are the most important ones in designing the sizing system since there are a lot of variations within the sample. By considering body dimensions with SD or coefficient of variations, we are actually considering the many different variations that exist in the sample of children. These variations should be the key element in creating a sizing system that will give good fit to the body. Nevertheless, not all these measurements can be used as the key dimensions as some of them are not practical or easily recognized by the consumer. Some of the significant and common key variables used for sizing system by the clothing experts and in ISO standards are bust girth, chest girth, waist girth, hip girth, and height [66–72].

In this section, the sample is grouped by body height and weight as these two body dimensions are considered to be the most distinctive trait for differentiating body shapes and types [73,74]. Very often body sizes are referred to height and weight of the human body. This is supported by Petrovia [57] who noted that the first trait by which objects can be distinguished is their largest dimensions—that is, height (tall or short) and weight (wide or narrow). Following is the analysis of height and weight of the measured sample.

4.6.2 Frequency distributions of data according to different gender

In this section, the dataset is analyzed by looking at the key body dimensions and its distribution. The frequencies of the important variables are reflected in the histograms for male and female gender, showing the critical dimensions of the anthropometric variables, namely height, weight, chest girth, hip girth, and waist girth [75]. A normal distribution line is superimposed over the histogram, providing an easy comparison to see if the data are skewed or bimodal.

4.6.2.1 Frequency distributions of data for the male sample

Figs. 4.10–4.14 are the frequency distributions of height, weight, chest girth, waist girth, and hip girth, respectively, for male samples with age range from 7 to 17 years old. These are the variables commonly used in the manufacturing of apparel. As can be seen in Fig. 4.10, the distribution of height for male samples is still showing bimodality. Further investigation still needs to be conducted to understand the cause of bimodality of the height dimensions. There are two main strata group for this sample population, which are the gender and age group. Since dividing the sample population according to gender does not reduce the bimodalism, thus the sample needs to be divided again according to gender and age group for investigation (as in the next section). On the other hand, weight distribution showed normal distribution and is the most variable. However, weight distribution is most skewed to the right followed by hip as the least skewed. In addition, weight is not comparable directly in the variation since it is measured in kg.

From the male samples (7–17 years old), referring to the figures below (Figs. 4.10–4.14), they showed that the distribution of height falls in the range of 101.0–184.4 cm. Each different sample population falls under one category of height, which can be seen in Fig. 4.10. The weight distribution is 14.6–146.7 kg. The chest for this group ranges from 35.3 to 134.0 cm. The minimum waist girth is 23.5 cm and the maximum is 124.2 cm. The minimum hip girth is 32.5 cm and maximum is 131.2 cm.

4.6.2.2 Frequency distributions of data for the female sample

In this section, Figs. 4.15–4.19 show five of the frequency distributions for females to gain some idea of the amount of scatter about the average. The distribution of females by stature is again bimodal and is less variable than either girth shown in the chart.

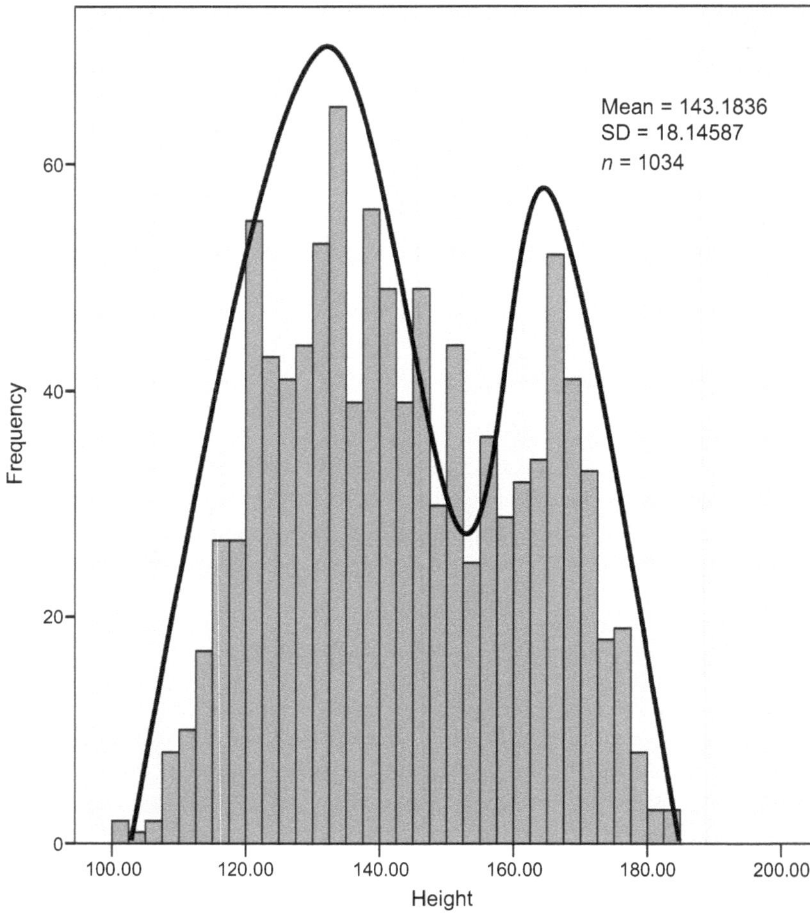

Figure 4.10 Frequency distributions of height of 1034 males.

Weight is most skewed to the right and chest and hip the least. Weight is not comparable to the other variables as the measurements were taken in kg. The minimum height for this population is 101.6 cm and the maximum height is 166.6 cm. The weight range is from 13.4 and 95.5 kg. The chest measurement for this group ranges from 51.5 to 114.5 cm. On the other hand, the range of hip girth is 47–121.5 cm. Lastly, the waist girth is distributed between 44 and 117 cm.

The distribution of key dimensions as shown in this chapter is useful for the estimation of size roll for the sizing system. Overall, the size range for height is 83.3 cm. As for weight, Otieno [59] has mentioned in her study that this dimension is useful only in smaller children. Previous studies clearly suggest that bust and hip girth should be used as key dimensions to replace weight. The reason for this is that both these dimensions (bust and hip girth) are flexible for varying size relationships for upper and lower body [54,58]. Another reason for substitution of weight is that

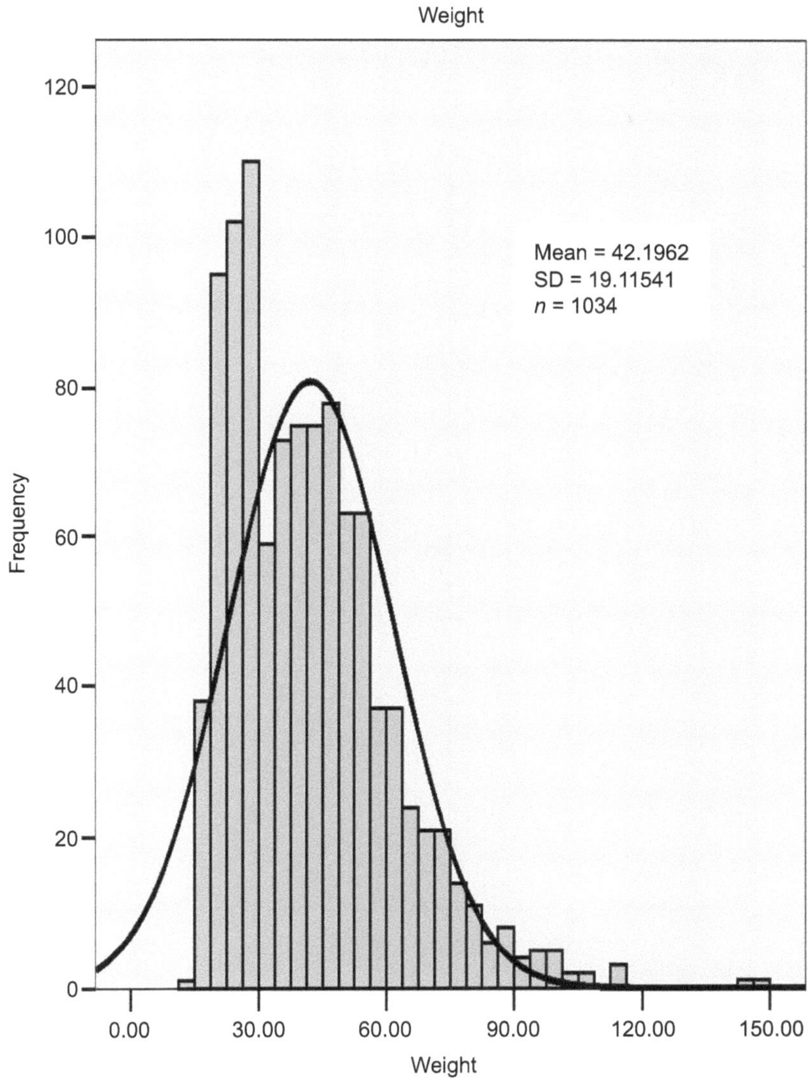

Figure 4.11 Frequency distributions of weight of 1034 males.

these dimensions (bust and hip girth) are the body dimensions in contact directly with the clothes worn while weight is not [55].

For the upper body, the size ranges for chest and bust girth are 98.7 and 89.0 cm, respectively. For the lower body, waist and hip girth measurements indicate size ranges of 101.0 and 98.7 cm, respectively. However, these extreme values seem high for size range. This indicates that the extreme values of the key dimensions may not be suitable for the size range; instead, the percentile value is more suitable. This is supported by previous studies in which the percentile value of key dimensions was used covering

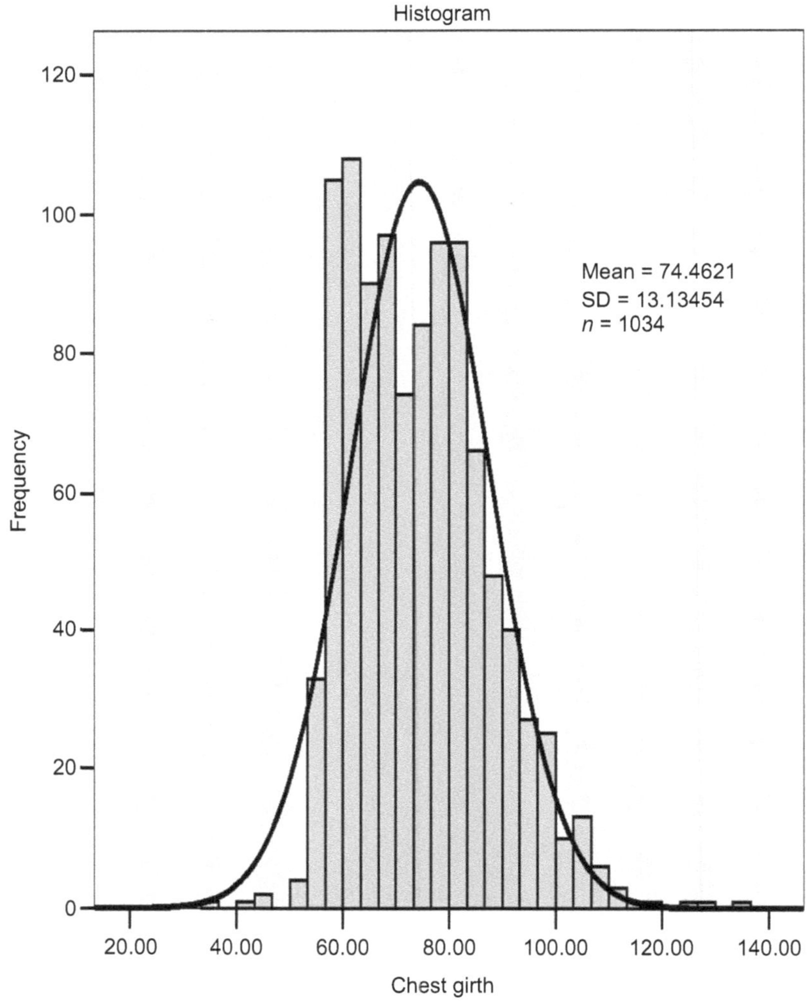

Figure 4.12 Frequency distributions of chest girth of 1034 males.

95% of the population [56]. Thus, for this study the percentile values are applied to the key dimensions where the 5th and 95th percentiles are used to accommodate 90% of the sample population. This percentile value will be applied to the key dimensions selected in the next stage of this study.

4.6.3 Independent samples t-test for key dimensions according to different gender

An independent samples *t*-test is conducted to compare the mean of each variable between male and female subjects. Referring to Table 4.6, there are significant differences in body dimensions, namely height, weight, waist girth, and hip girth

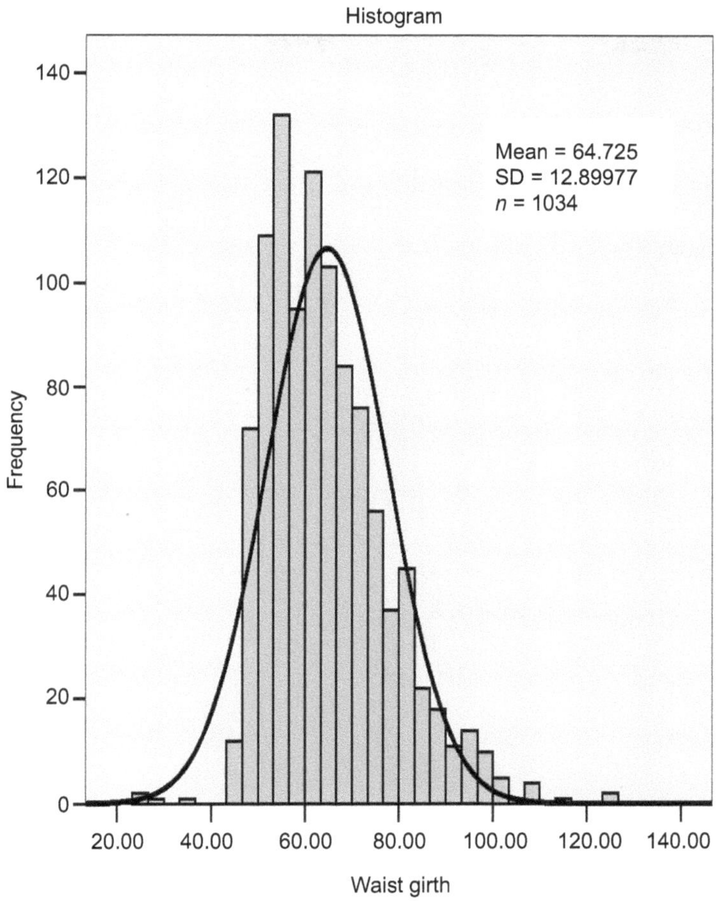

Figure 4.13 Frequency distributions of waist girth of 1034 males.

between males and females. However, chest dimensions showed no differences when compared between males and females. The results also show that males have higher mean anthropometric measurements than females (ie, males are in general larger than females).

This finding is in agreement with previous studies that show males are bigger than females except in bust and hip girth [15,60]. In this section, only the results of key dimensions (height, weight, chest girth, waist girth, and hip girth), as identified earlier, are presented and discussed.

From Table 4.6, significant t-test shows that males are overall taller ($M = 143.2$ SD$=18.2$) than females ($M=140.6$, SD$=15.1$) with $t(3.49) = 36.6$, $p < 0.001$. Males are also heavier ($M=42.2$ SD$=19.1$) than females ($M = 39.32$, SD $= 13.1$) with $t(3.8)=33.55$, $p<0.000$. In addition, when comparing chest for both genders, chest was not significantly different between males ($M = 74.5$ SD $= 13.1$) and females ($M=73.1$, SD $= 12.7$) in general showing $t(1.5) = 23.2$, $p>0.001$. Next, waist girth for

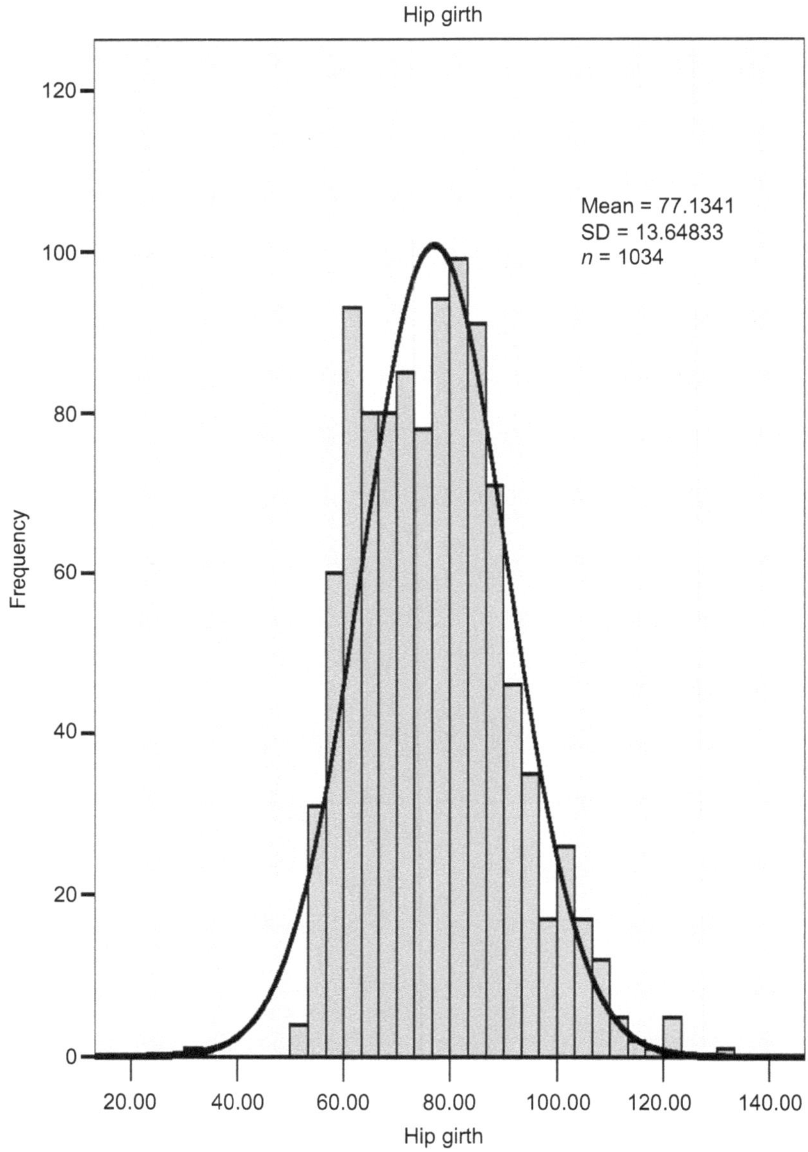

Figure 4.14 Frequency distributions of hip girth of 1034 males.

males ($M=64.7$ SD $=12.9$) and females ($M=62.8$, SD $=10.7$) are significantly different, being larger in males with $t(3.61)=32.7$, $p<0.001$. Lastly, hip girth measurement is significantly different, being larger in females ($M=79.3$, SD $=13.4$) as compared to males ($M=77.1$, SD $=13.6$), with $t(-3.64)=0.06$, $p<0.0001$.

From Fig. 4.20 it can be observed that all the key dimensions mentioned herein are considerably different between genders except chest. There is no significant difference in chest girth between males and females, making it important to take bust

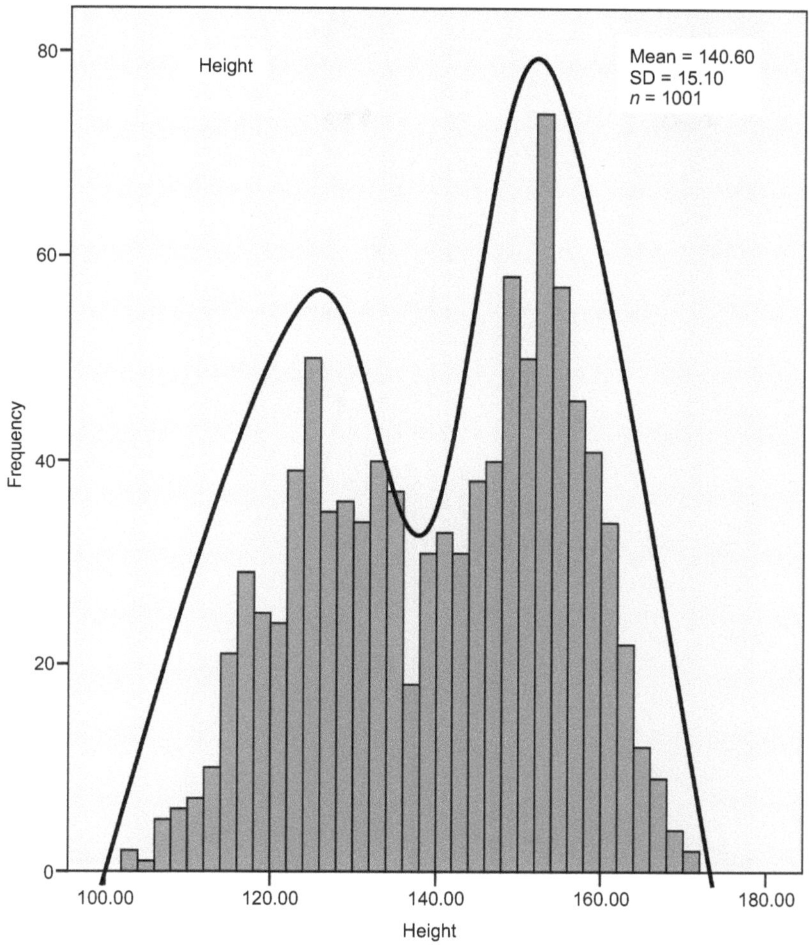

Figure 4.15 Frequency distributions of height of 1001 females.

measurements for females as compared to chest girth. The key dimensions showed significant differences between genders, which substantiate studies in the literature suggesting differences of body dimensions between males and females [15]. In addition, the present findings seem to be consistent with research by Chung et al. [15] and Ariadurai et al. [76] who both developed sizing systems for males and females (children) separately due to these differences. Specifically, this study confirms that differences in body dimensions do indeed exist between different genders.

As said in the earlier paragraphs, the bimodalism of height distribution required further investigation. Referring to Fig. 4.4, the height distribution for the sample size shows bimodality distribution. This bimodality trend is found in height probably due to the existence of different genders in the sample size. After splitting the sample size according to male and female as shown in Figs. 4.9 and 4.11, it is interesting to see that the two peaks still exist although the samples are separated according to gender.

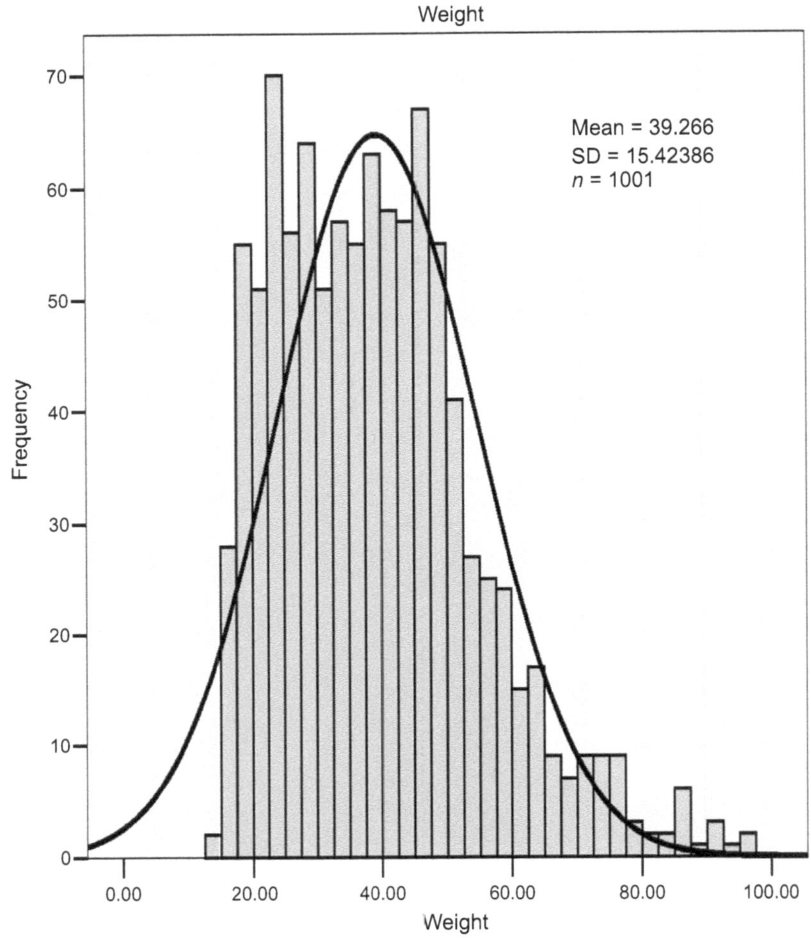

Figure 4.16 Frequency distributions of weight of 1001 females.

The bimodalism must be interpreted with caution according to Schilling et al. [48] because in some cases height distribution may look bimodal but in actuality it is not.

Let us analyze the true meaning of bimodal characteristics in histogram graphs. To be truly bimodal the following condition must be met: the difference in both means must differ by more than approximately the sum of their standard deviations. However, in this study the results did not meet this criterion; the differences of mean height for both genders were 2.6, which is less than the total SD of 33.3. Therefore, it can be concluded that the bimodalism in height distribution is not due to gender differences as is obvious from Figs. 4.9, 4.11, and 4.21.

Further analysis to explain the bimodalism is done by splitting the samples into two age ranges, namely ages 7–12 (primary school age groups) and ages 13–17 (secondary school age groups). The splitting according to age is done to investigate whether the two distinct groups still exist in the distribution. As shown in Fig. 4.22, after splitting the samples according to two different age groups, the bimodalism vanishes.

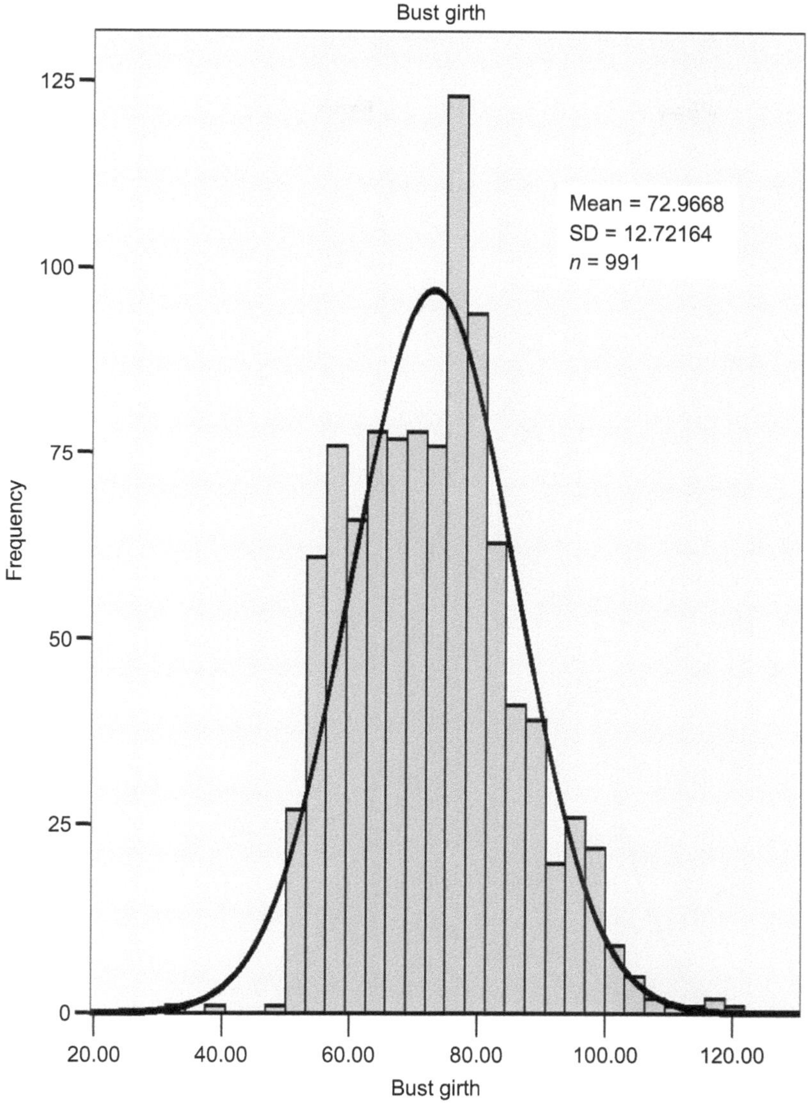

Figure 4.17 Frequency distributions of bust girth of 991 females.

The graph looks symmetrical, which confirms that there is a distinct age group within the samples. Therefore, it can be concluded that the peak measurements from Fig. 4.9 belong to the different age group as shown in Fig. 4.10, and therefore is the cause of the bimodalism.

The same investigation is conducted on the female samples with the same results (see Figs. 4.23 and 4.24). Thus, this demonstrates the two peaks (bimodalism) exist because two distinct age groups represent the sample heights. As a conclusion, the sample size is finally split into four distinct groups based on gender and age groups as seen in Figs. 4.22 and 4.23 for further analysis in the next stage, which is the sizing analysis.

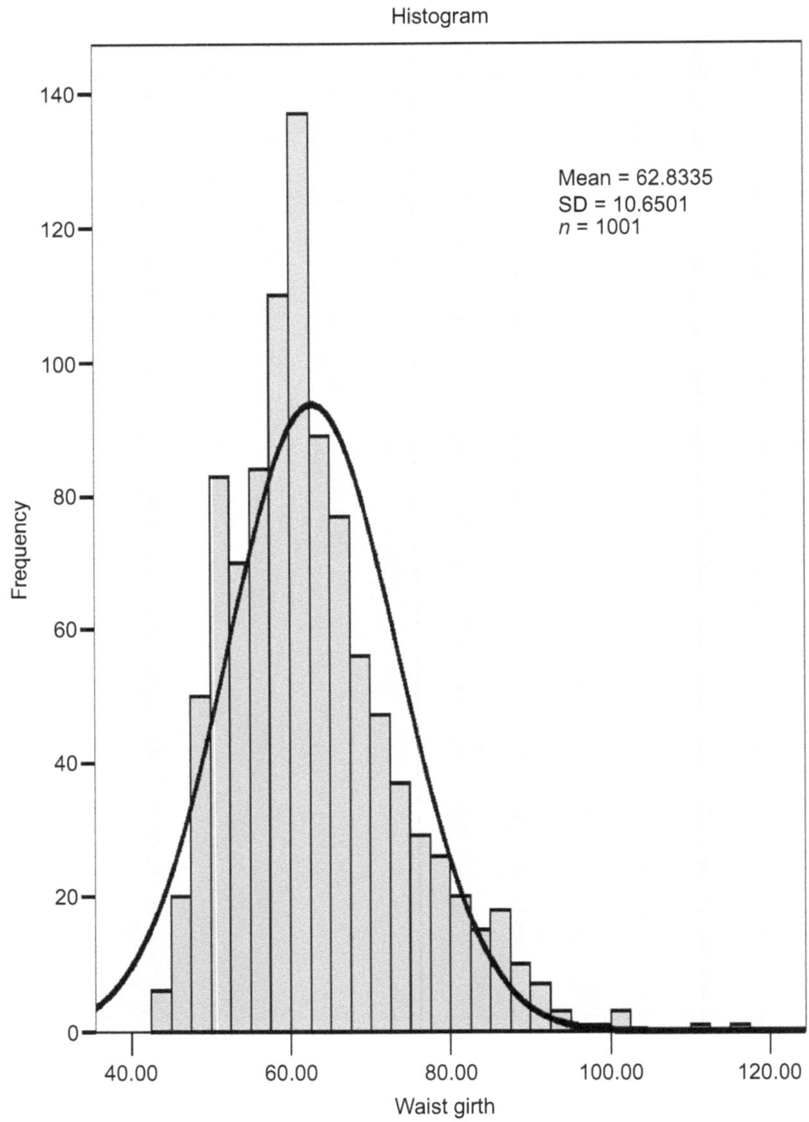

Figure 4.18 Frequency distributions of waist girth of 1001 females.

4.7 Continuous data results and analysis for sample population according to gender and age

For this section, the mean results for all the anthropometric dimensions are tabulated according to age and gender. The variations of 11 sizes for ages 7–17 are carefully analyzed. The objective of this section is to test whether the means of body dimensions differ significantly between age and gender using MANOVA test. Only selected key dimensions

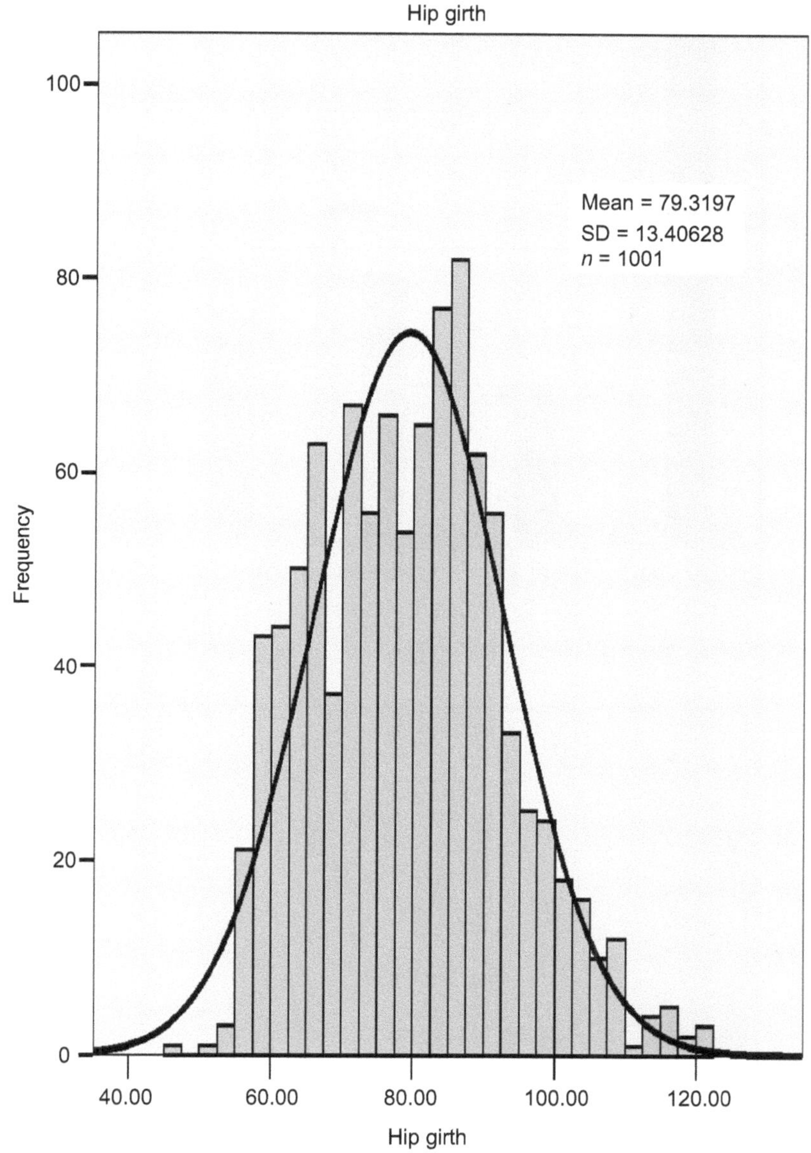

Figure 4.19 Frequency distributions of hip girth of 1001 females.

Table 4.6 Independent *t*-test for differences between genders for key body dimensions

Statistic	*F*	*t*	*df*	Significant	Prob
Height	36.578	3.475	2033	Significant	0.001
Weight	33.554	3.798	2033	Significant	0.000
Chest girth	23.211	1.501	2033	Not significant	0.133
Waist girth	32.664	3.601	2033	Significant	0.000
Hip girth	0.062	−3.643	2033	Significant	0.000

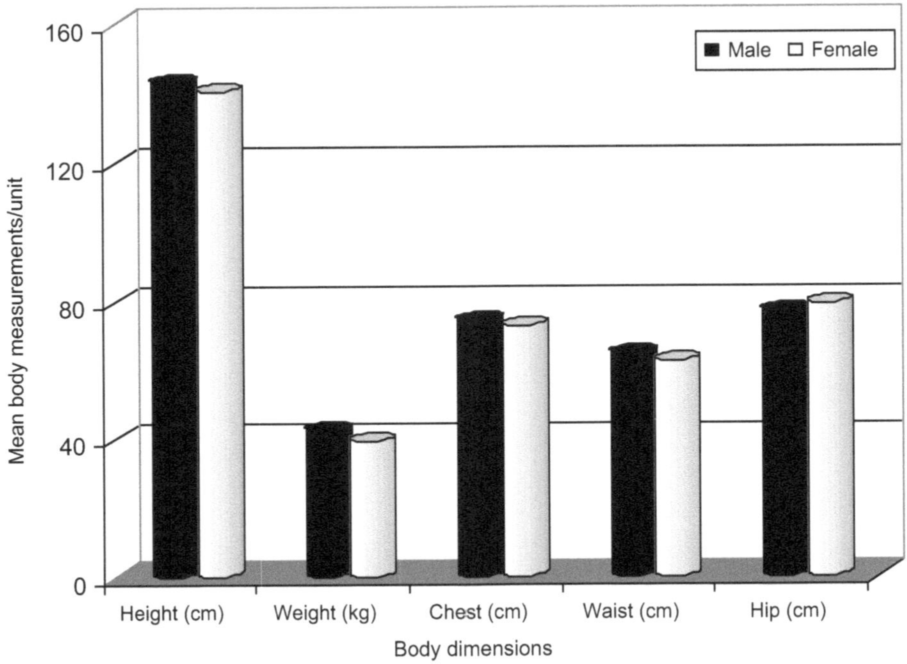

Figure 4.20 Mean key body dimensions for males and females ages 7–17.

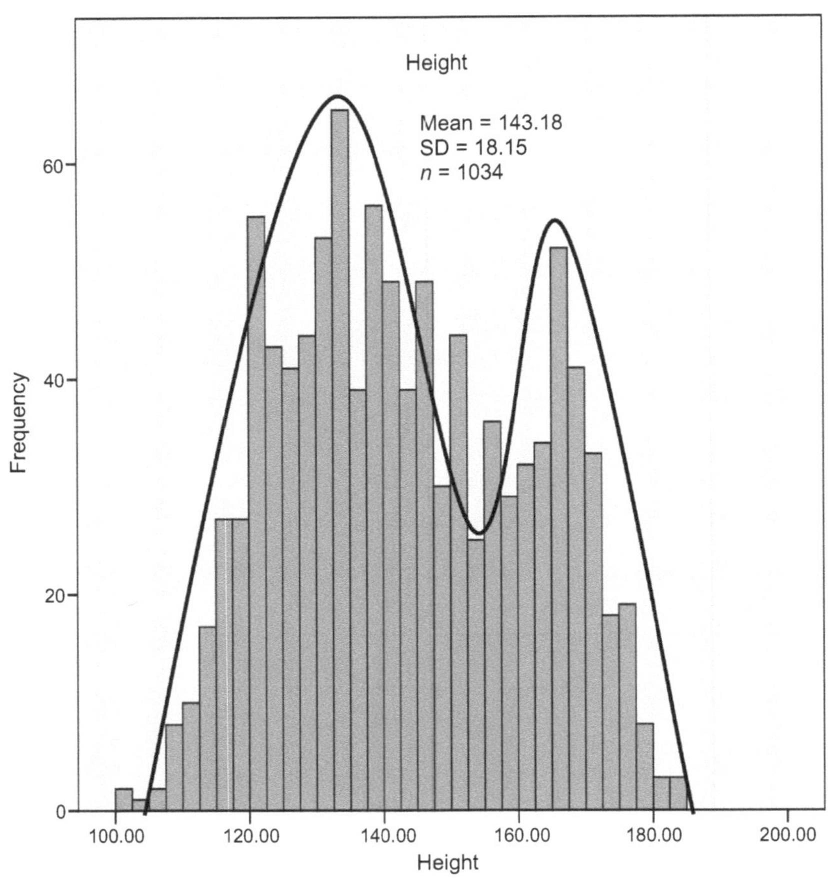

Figure 4.21 Frequency distributions of height for males ages 7–17 years old.

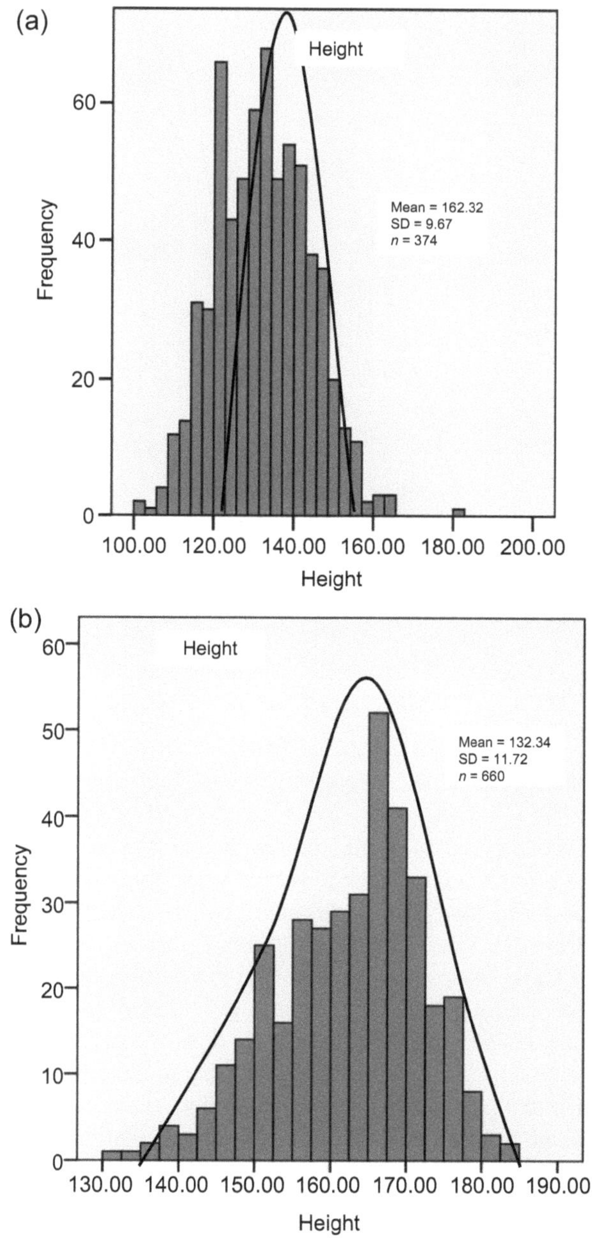

Figure 4.22 Frequency distributions of height for males (a) 7–12 years old and (b) 13–17 years old.

are used. In general, the results of this study show that there are significant differences in key body dimensions of height, bust girth, chest girth, waist girth, and hip girth according to gender and age. The majority of the dimensions are higher in female subjects during the age range of 7–12 years old. Toward adolescence (from 13 years old), significant gender differences are observed in the dimensions with larger measurements in the male subjects.

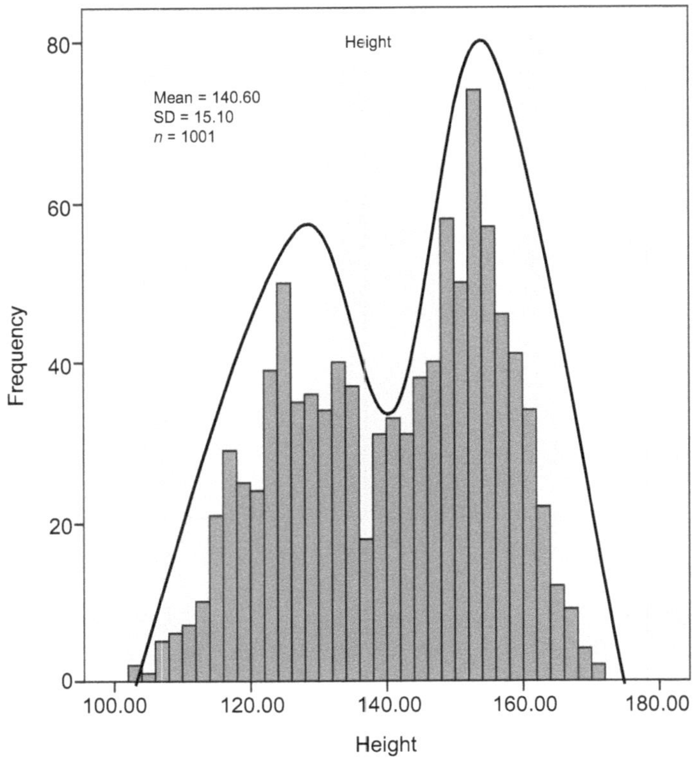

Figure 4.23 Frequency distributions of height for females 7–17 years old.

The values of Wilks's lambda in this study are reported to be in the range of 0.2–0.3. Wilks's lambda ranges from 0 to 1; the lower the value, the more the given effect contributes to the model. The MANOVA test results (see Table 4.7) show $p<0.000$, which is significant at alpha = 0.05. Thus, significant differences in these key body measurements exist among the 11 age groups and genders. The lower value of lambda indicates that the key dimensions were extremely effective in differentiating the 11 age groups and genders. In addition, it also reveals that these body dimensions differ by age and gender.

These results also substantiate previous studies that found differences in key dimensions based on age and gender [15,39,76]. In addition, the effect of gender is proven by a study carried out by Barli et al. [11] who identified significant gender differences in anthropometrical measurements.

One possible reason why growth is similar in both genders from ages 7 to 9 is that this is the phase where rapid growth slows down temporarily before it takes off again during puberty (10–13 in females and 11–14 in males). The growth spurt at puberty also explains why at ages 10–12 females show bigger differences as compared to males, and after age 13 the males are bigger. Overall, two important findings of this study are:

1. It implies that the growth of body dimensions is highly dependent on age and gender.
2. Body measurements are bigger in different genders at different ages. This finding is supported by previous literature.

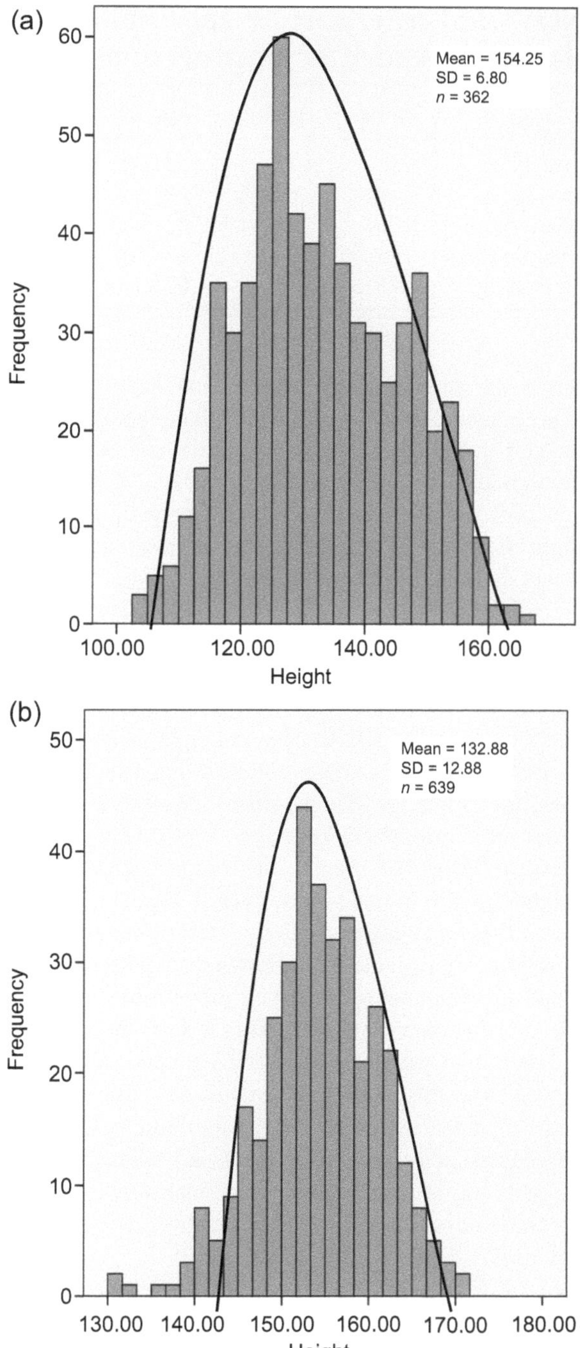

Figure 4.24 Frequency distributions of height for females (a) 7–12 years old and (b) 13–17 years old.

Table 4.7 **MANOVA test criteria and F approximation for differences of age and gender for key body dimensions**

Statistic	Value	*F* value	Num DF	DenF	PrF
Wilks's lambda (height)	0.107	4.77	1224.000	2844.000	0.000
Wilks's lambda (weight)	0.191	2.68	1322.000	2746.000	0.000
Wilks's lambda (chest girth)	0.264	3.56	854.000	3212.000	0.000
Wilks's lambda (waist girth)	0.332	2.91	854.000	3212.000	0.000
Wilks's lambda (hip girth)	0.257	3.18	952.000	3112.000	0.000

In the next section, the samples are grouped by body height and weight as these two body dimensions are considered to be the most distinctive trait for differentiating body shapes and types [72,77]. Very often, body sizes are referred to in terms of height and weight of the human body. This is supported by Petrovia [63] who noted that the first trait by which objects can be distinguished is their largest dimension that is height (tall or short) and weight (heavy or light). The results of height and weight analysis of the measured sample are discussed in the next section.

4.7.1 Height distribution according to gender and age

Fig. 4.25 shows the height distributions of male and female subjects. It illustrates the height changes from age 7 to 17. Differences in height growth are observed between different genders. Both males and females follow a linear increase in height from age 7 to 12, after which the growth for females slows and stops at 156 cm.

The females are a little bit taller during ages 7 to 9. However, different growth is observed in females from 7 to 14 at 35 cm, while for males growth is 48 cm from 7 to 15, which indicates higher growth in males. However, it is earlier in females. This may be explained by the fact that females enter puberty earlier than males, between ages 8 to 13 (243–245). In another study, it has been shown that females start their growth spurt between ages 10 and 12 while males begin their growth spurt between ages 12 and 14, although it can be later in some cases [25]. This is supported by this study, in which males were seen to continue growing from age 13 onward.

It is also observed from this study that females are taller than the males between the ages of 10 and 12. This is supported by the literature, which states that from age 11 or 12 to adolescence (known as puberty) girls are taller than boys. In turn, as the boys reach age 13 or 14, during the puberty years, males increase in height faster than females. This is because boys start to have their major growth spurt sometime between the ages of 10 and 15 [78,79].

From the age of 14, girls' height starts to show slow growth. After the age of 14, the female height seems to become stagnant as compared to male height. This is because the female's growth associated with puberty has ended and sexual maturity has been reached; as a result, height is relatively the same for females from age 15 to 17 [14,80].

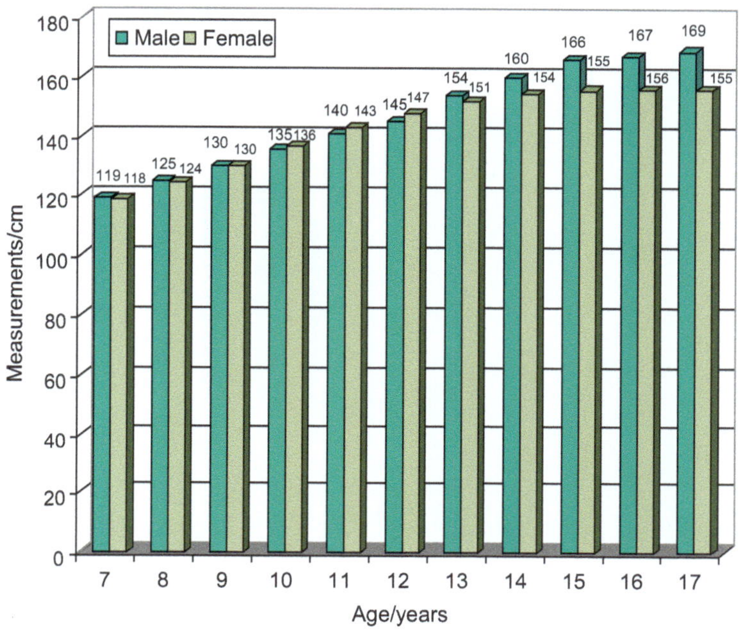

Figure 4.25 Height distributions according to gender and age groups.

On the other hand, the male sample's height continues to grow until age 17 where their height reaches 169 cm. Males and females have about the same mean height from 7 to 9 years old. Female mean height increased rapidly from 118 to 154 cm from age 7 to 14, an increase of 36 cm. From age 14 to 17, the female mean height showed much slower growth, from 155 to 156 cm, an increase of only 1 cm. As for males, they show a similar pattern. From age 7 to 15 the mean height showed a significant change from 119 to 166 cm, an increase of 47 cm. From age 15 to 17, height increase was slow, from 167 to 169 cm, an increase of 2.8 cm.

In conclusion, clear height differences are observed between genders. Thus, the separation of the sizing system according to height for each gender is possible and important because garment length depends on the height of individuals. This also implies that a dynamic sizing system should be designed according to different age groups since the age shows significant differences.

As mentioned in the earlier analysis, height is strongly correlated with age ($r=0.86$), which means that as age increases height also increases. According to Gupta et al. [10], a proportional relationship among body measurements is significant in a sizing system as these dimensions can be selected as key dimensions. However, although height and age are strongly correlated as shown in Fig. 4.26, it is obvious in every age group that the range of height is spread out from 101 to 130 cm for 7 years old. For example, in age 7 alone, the variation of height is 29 cm. The same trend of spread is seen in all ages. Not only do children of the same age have widely different heights but the same height measurement can be found in widely

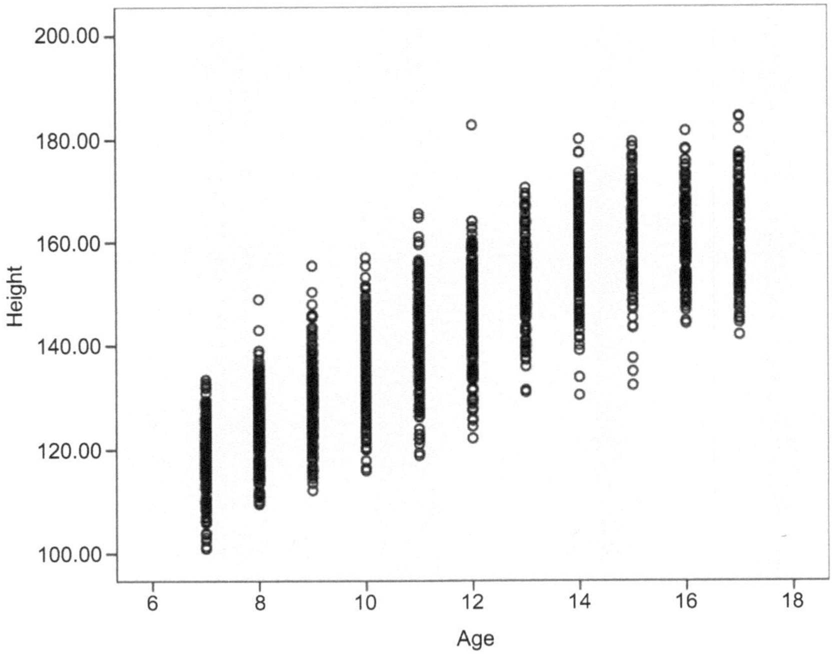

Figure 4.26 Scatter plot of height versus age according to gender.

different ages. Thus, age is not suitable as a factor for size system development because age is not the determining factor for height. The same finding is seen in other sizing studies [39,53,81].

Moreover, the spread of height is higher between ages 7 to 12 as compared to ages 13 to 17. This finding is consistent with those of A'Hearn et al. [51] who found that differences in height by age are the largest between the ages of 9 and 13. Other studies also show that height is affected by age, nutrition, lifestyle, and health [59]. It can, therefore, be assumed that there will be more sizes needed for ages 7–12 as compared to ages 13–17 due to the larger size range in the younger group. Another possible reason for a lower size range for ages 13–17 is their slow height growth as compared to ages 7–12. However, the result of a higher size roll in ages 7–12 as compared to ages 13–17 depends on the size interval and considerations of the spread of the population within that age range. It is the goal of any sizing table to accommodate as many people as possible.

4.7.2 Weight distribution according to gender and age

The mean weight distribution for female and male subjects is presented in Fig. 4.27. This graph illustrates the differences of weight distribution among different genders. MANOVA test (Table 4.7) indicates that weight shows statistically significant differences between age and gender. Overall, the mean weight increased substantially for

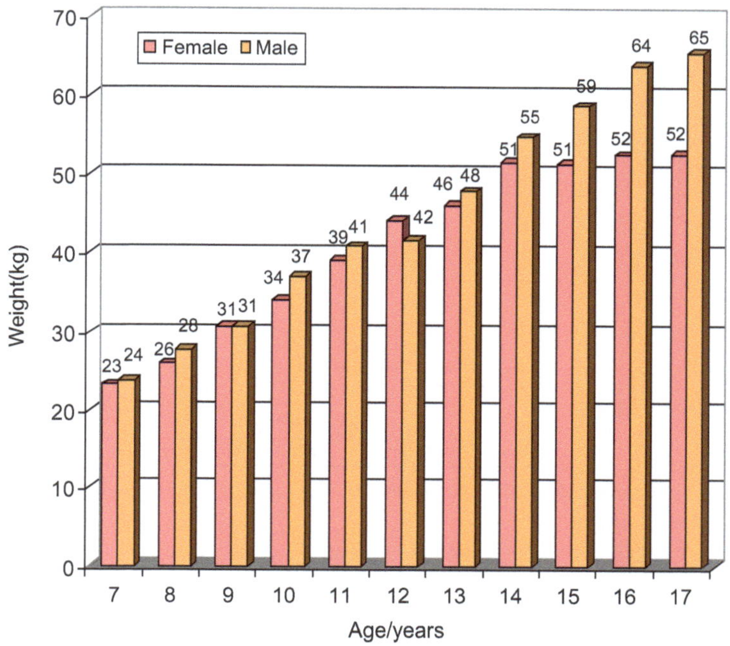

Figure 4.27 Distribution of weight according to gender and age group.

both females and males ages 7–17, with two different growth patterns in females and three different growth patterns in males.

For females, weight shows a significant increment from age 7 to 14. Then from age 15 to 17, weight shows very little increment (1 kg). Males show a significant increment of weight from age 7 to 11. Between 11 and 12, a very small difference of weight is noted (1 kg). After age 12, weight shoots up until age 16, and then from 16 to 17, weight stabilizes with little difference of increment (1 kg).

Overall, the total increment of weight is larger in males than in females. At ages 7–17, the mean weight for males is 24–66 kg, an increase of 42 kg. In contrast, the female samples show a mean weight of 23–52 kg, an increase of only 29 kg. It is also quite clear from Fig. 4.23 that the male's weight is higher than the female's at every age with the exception of age 12.

4.7.3 Girth body dimensions according to males and age

Fig. 4.28 shows the measurements of girth (horizontal) body dimensions of male samples for ages 7–17. In general, children go through three phases of growth. At age 7, measurements for chest, hip, and waist girth start at 63, 65, and 55 cm, respectively. As they grow toward puberty at about age 12, there is a steady rise from 2 to 5 cm per year. There is a transition period of very slow growth at age 12 or 13. After age 13, the line resumed increasing, but at a slower rate as compared to ages 7–12. The total increment

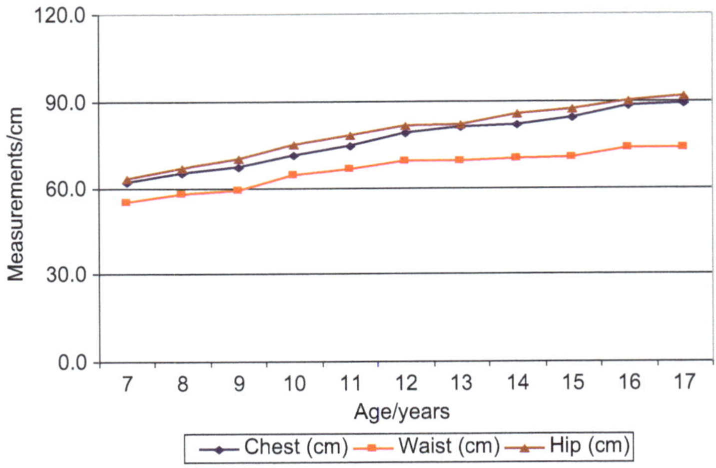

Figure 4.28 Comparison of girth dimensions for chest, waist, and hip girth among 7- to 17-year-old male children.

found in chest, waist, and hip girth for ages 7–12 is 17.1, 14.1, and 18 cm, respectively. As for the increment at ages 13–17, for chest, waist, and hip girth the numbers are 8.1, 4.4, and 9.6 cm, respectively.

Obviously, these data suggest that there is an increment of measurements in all ages with the exception between age 12 and 13 in chest and hip girth and for 12 and 14 in waist girth. Overall, there is a continuous increment of measurements with age. This pattern of growth also points out that there is a transition period of growth from rapid (7–11) to stagnant (12–13) to slow (13–17). Therefore, it is rational to divide this age range by the two important age groups of 7–12 and 13–17. Another rationale for this group segregation is the bimodalism of height distribution (see Section 4.5 for discussion of bimodalism).

Table 4.8 shows the results of ANOVA test. The one-way ANOVA test is conducted to see if the growth increments of dimensions mentioned in Fig. 4.29 are significantly different among the ages for male samples. The results of ANOVA test show significant age differences for all key body dimensions including height and weight. Based on the ANOVA results in Table 4.8, at a significance level of 0.05, the mean body dimensions for the 11 age groups are significantly different with $p<0.001$. The result for height $F(10,1023)=463.54$, weight $F(10,1023)=93.00$, chest girth $F(10,1023)=72.83$, waist girth $F(10,1023)=26.69$, and hip girth $F(10,1023)=70.67$ all show significant $p=0.000$. This result implies that the increment is significant and thus the growth of each key dimension is found to be significant.

Next, one-way ANOVA testing is conducted to look at the differences of five key body measurements for two different age groups. Fig. 4.29 shows the growth trends of the two age groups. Based on the ANOVA results (see Table 4.9), at a significance level of 0.05, it shows that the mean body dimensions for the two age groups are significantly different with $p<0.001$. The results for height $F(10,1032)=1765.9$, weight $F(1,1032)=574.4$, chest girth $F(10,1032)=428.4$, waist girth $F(10,1032)=141.8$, and hip girth $F(10,1032)=397.7$ are all showing significant value with $p=0.000$.

Table 4.8 ANOVA table for growth differences male samples

ANOVA						
		Sum of squares	***df***	**Mean square**	**F**	**Sig.**
Height	Between groups	278,644.0	10	27,864.402	463.541	0.000
	Within groups	61,494.602	1023	60.112		
	Total	340,138.6	1033			
Weight	Between groups	179,738.8	10	17,973.879	92.997	0.000
	Within groups	197,718.1	1023	193.273		
	Total	377,456.9	1033			
Chest girth	Between groups	74,112.352	10	7411.235	72.833	0.000
	Within groups	104,096.8	1023	101.756		
	Total	178,209.1	1033			
Waist girth	Between groups	35,572.025	10	3557.203	26.694	0.000
	Within groups	136,323.3	1023	133.258		
	Total	171,895.3	1033			
Hip girth	Between groups	78,623.150	10	7862.315	70.677	0.000
	Within groups	113,800.9	1023	111.242		
	Total	192,424.0	1033			

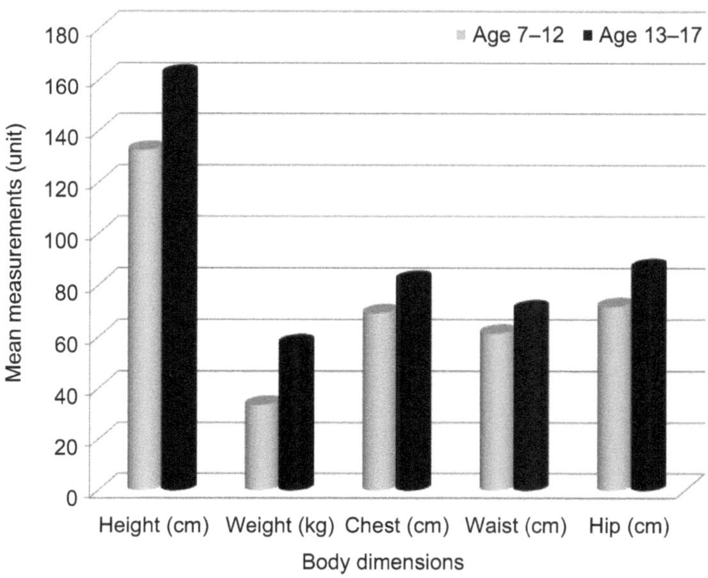

Figure 4.29 Comparison of growth trend between two age groups for males.

This test proves that the two age groups differ significantly according to all the body dimensions, including height, weight, chest girth, waist girth, and hip girth. The test also justifies the classification into two age groups for the development of a sizing system.

Table 4.9 **ANOVA table for male samples according to different age group**

ANOVA						
		Sum of squares	***df***	**Mean square**	***F***	**Sig**
Height	Between groups	214,679.9	1	214,679.906	1765.917	0.000
	Within groups	125,458.7	1032	121.569		
	Total	340,138.6	1033			
Weight	Between groups	134,960.5	1	134,960.508	574.356	0.000
	Within groups	242,496.4	1032	234.977		
	Total	377,456.9	1033			
Chest girth	Between groups	52,271.397	1	52,271.397	428.339	0.000
	Within groups	125,937.7	1032	122.033		
	Total	178,209.1	1033			
Waist girth	Between groups	20,760.882	1	20,760.882	141.763	0.000
	Within groups	151,134.4	1032	146.448		
	Total	171,895.3	1033			
Hip girth	Between groups	53,523.359	1	53,523.359	397.666	0.000
	Within groups	138,900.7	1032	134.594		
	Total	192424.0	1033			

4.7.4 Girth body dimensions according to female and age

Fig. 4.30 shows the growth of horizontal body measurements for females ages 7–17. The graph shows the transitions of growth according to three obvious phases. Overall, there is a continuous increment of measurements for all body dimensions, which shows that as age increases, girth body measurements also increase. However, the increment rate tends to differ from age 7 to 12, 12 to 13, and 13 to 17. There is a rapid growth from age 7 to 12, then the growth seem to slow down from age 12 to 13, and then increase slightly from age 13 to 17. Another obvious observation from the graph is the slow growth from age 14 to 17, during which it is noted that all girth measurements seem to stabilize without much increment. This results in the total growth of key body measurements for age 13 to 17 being much lower than that of age 7 to 12.

Moreover, the body dimensions showed rapid growth from age 7 to 12 with total increments of 17.5, 12.3, and 18.3 cm for bust, waist, and hip girth, respectively. In comparison, the slower increments in bust, waist, and hip girth are 5.1, 2.1, and 5.7 cm, as can be seen in the measurements from age 13 to 17.

Based on Table 4.10 for ANOVA results, at a significance level of 0.05, the mean body dimensions for the 11 age groups (ages 7–17) are significantly different with $p<0.001$. The results for height $F(10,990)=298.98$, weight $F(10,990)=82.95$, bust girth $F(10,990)=79.10$, waist girth $F(10,990)=25.85$, and hip girth $F(10,990)=95.79$ are all showing significant $p=0.000$. This result implies that the increment is significant

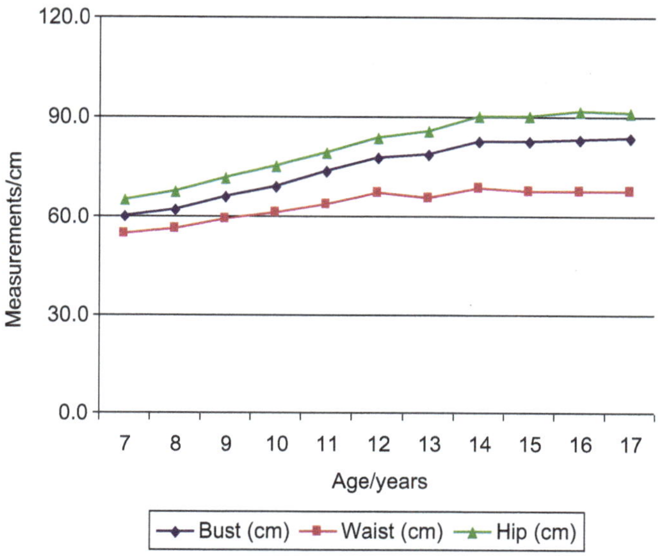

Figure 4.30 Comparison of girth dimensions for bust, waist, and hip girth among 7–17 year old female children.

Table 4.10 ANOVA table for growth differences in female samples

ANOVA						
		Sum of squares	***df***	**Mean square**	***F***	**Sig.**
Height	Between groups	171,359.1	10	17,135.906	298.984	0.000
	Within groups	56,740.684	990	57.314		
	Total	228,099.7	1000			
Weight	Between groups	108,455.7	10	10,845.573	82.951	0.000
	Within groups	129,439.8	990	130.747		
	Total	237,895.5	1000			
Bust girth	Between groups	71,711.280	10	7171.128	79.097	0.000
	Within groups	89,756.269	990	90.663		
	Total	161,467.5	1000			
Waist girth	Between groups	23,483.801	10	2348.380	25.849	0.000
	Within groups	89,940.747	990	90.849		
	Total	113,424.5	1000			
Hip girth	Between groups	88,384.756	10	8838.476	95.793	0.000
	Within groups	91,343.606	990	92.266		
	Total	179,728.4	1000			

Table 4.11 ANOVA table for female samples according to different age group

ANOVA						
		Sum of squares	***df***	**Mean square**	***F***	**Sig.**
Height	Between groups	101,354.9	1	101,354.945	798.878	0.000
	Within groups	126,744.8	999	126.872		
	Total	228,099.7	1000			
Weight	Between groups	69,093.950	1	69,093.950	408.911	0.000
	Within groups	168,801.5	999	168.971		
	Total	237,895.5	1000			
Bust girth	Between groups	44,876.883	1	44,876.883	384.525	0.000
	Within groups	116,590.7	999	116.707		
	Total	161,467.5	1000			
Waist girth	Between groups	10,278.723	1	10,278.723	99.553	0.000
	Within groups	103,145.8	999	103.249		
	Total	113,424.5	1000			
Hip girth	Between groups	57,313.785	1	57,313.785	467.726	0.000
	Within groups	122,414.6	999	122.537		
	Total	179,728.4	1000			

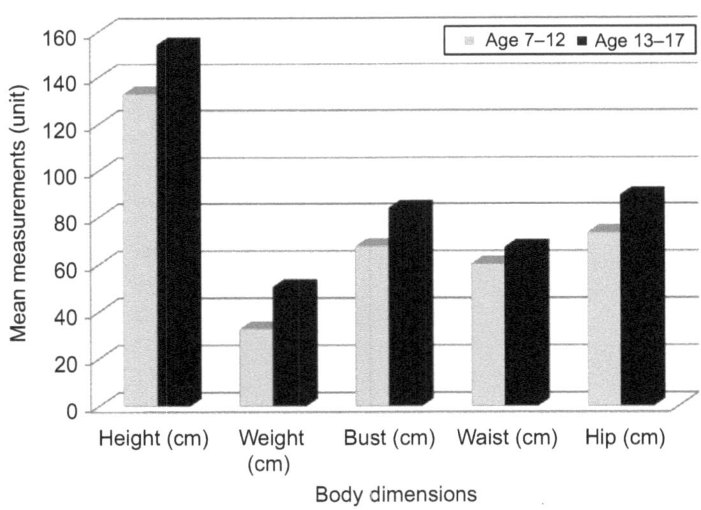

Figure 4.31 Comparison of female growth trend between two age groups.

and thus confirms the finding of many previous studies that these dimensions are suitable for use as the key dimensions.

Looking at the results of girth dimensions in female samples, there is a possibility to divide the group into two distinct groups, 7–12 and 13–17 years old. Thus, one-way ANOVA test shown in Table 4.11 is conducted again to check the significant differences of both age groups. Based on the ANOVA results, at a significance level of 0.05, it shows that the mean body dimensions for ages 7–12 and 13–17 are significantly different with $p<0.001$. The results for height $F(1,999)=798.88$, weight $F(1,999)=408.91$, bust girth $F(1,999)=384.53$, waist girth $F(1,999)=99.55$, and hip girth $F(1,999)=467.73$ are all showing significant $p=0.000$.

Next, key body measurements according to different age group are shown in Fig. 4.31. The figure shows the growth trends of key body dimensions according to the two age groups.

Thus, this test proves that both age groups differ significantly according to the key body dimensions, namely the height, bust girth, waist girth, and hip girth. The test also justifies the classification into two age groups for the development of a sizing system.

4.8 Conclusions

The results of this chapter show that anthropometric differences between genders exist. Significant age differences between anthropometric data are also revealed. Therefore, from this chapter it is confirmed that the sample can be divided into two genders (male and female) and two age groups (7–12 and 13–17). This finding substantiates the need to develop a sizing system according to different gender and age groups as suggested previously [14,15,82].

In addition, the findings from this chapter suggest that four dimensions—namely height, chest, bust, waist, and hip girth—can be selected as the key dimensions. These dimensions are highly correlated and thus can be regarded as significant variables by which to classify the sample population into homogenous groups for the development of a sizing system.

This chapter also proves that body dimensions are highly dependent on age and gender. Children from age 7 to 17 are found to have wide variations of height but different ages of children are also found within the same height measurement. This finding suggests that age is not appropriate as the determining factor for a size system but that the body dimensions are. It also shows that no specific height can be associated with a certain age, as humans come in many different sizes and shapes. This finding supports previous studies that concluded it is not appropriate to develop a sizing system for children according to age [15,58]. In addition, the spread of measurement is not only found in height and age but in girth measurements and length measurements. Therefore, it is essential that the combination of key dimensions come from girth and length measurements in order to fit the population well, as recommended by many researchers of previous studies [52,59,69,83]. Thus, this chapter concludes the general overview of the body sizes and shapes of the sample population, which represents the population of children in one country. This chapter shows how the analysis of the anthropometric data is conducted in order to understand the body sizes and shapes of the sample population, which is the first step before determining the key dimensions for the development of the sizing system for children's wear. The next chapter will be the data analysis to identify the key dimensions that will be used to cluster the sample population into similar traits of body dimensions.

References

[1] Z. Peter, Cluster analysis in stoichiometric research: a pattern-oriented approach to identifying temporally stable peer status groups of girls, J. Early Adolesc. 27 (2007) 90–114.
[2] H.L. Viktor, E. Paquet, H. Guo, Measuring to fit: virtual tailoring through cluster analysis and classification, in: Presented at 17th International European Conference on Machine Learning and the 10th European Conference on Principles and Practice of Knowledge Discovery in Database, September 18–22, 2006. Berlin, Germany.
[3] G. Cuvier, Le Regne Animal Distribue E'apres Son Organisation, Fortin, Paris, 1828.
[4] G. Darwin, On the Origin of the Species, 1859.
[5] J. Huxley, Problems of Relative Growth, 1932.
[6] S.R. Lele, J.T. Richstmeier, An Invariant Approach to Statistical Analysis of Shapes: Interdisciplinary Statistics, 2001.
[7] L.D. Cohn, N.E. Adler, C.E. Irwin, S.G. Millstein, S.M. Kegeles, G. Stone, Body-figure preferences in male and female adolescents, J. Abnorm. Psychol. 96 (1987) 276–279.
[8] P. Devarajan, C.L. Istook, Validation of female figure identification technique (FFIT) for apparel software, J. Text. Appar. Technol. Manag. 4 (2004).
[9] D. Singh, Body shape and women's attractiveness: the critical role of waist-to-hip ratio, Hum. Nat. 4 (1993) 297–321.
[10] D. Gupta, Indian Body Dimensions, Tex exchange (TC2), 2008.
[11] A. Petrova, S.P. Ashdown, Three-dimensional body scan data analysis: body size and shape dependence of ease values for pants' fit, Cloth. Text. Res. J. 26 (2008) 227–252.
[12] C.H. Hsu, Applying a bust-to-waist ratio approach to develop body measurement charts for improving female clothing manufacture, J. Chin. Instit. Ind. Eng. 25 (2008) 215–222.
[13] D. Mahadzir, In a fit over sizing, in: Sunday Star. 21st August, 2005. Kuala Lumpur.
[14] Y. Kang, H.S. Choi, W.H. Do, A study of the apparel sizing of children's wear. An analysis of the size increments utilized in children's wear based on an anthropometric survey, J. Korean Home Econ. Assoc. Engl. Ed. 2 (2001) 96–110.
[15] A.S. Ariadurai, T.P.G. Nilusha, M.R. Dissanayake, An anthropometric study on Sri Lankan school children for developing clothing sizes, J. Soc. Sci. 19 (2009) 51–56.
[16] A.M. Mason, D. Klerk, S.P. Ashdown, Sizing and fit research at grassroots level. A methodology for the identification of unique body shapes in African developing countries, J. Fam. Ecol. Consume. Sci. 36 (2008) 9–21.
[17] C.H. Hsu, H.-F. Lin, M.J.J. Wang, Y.C. Lin, An application of data mining technique in developing sizing system for army soldiers in Taiwan, WSEAS Trans. Comput. Arch. 7 (2008) 245–252.
[18] S.R. Daniels, Body shape in children and adolescents, J. Pediatr. 152 (2008) 45.
[19] T.J. Eveleth, Worldwide Variation in Human Growth, second ed., Cambridge University Press, Cambridge, 1990.
[20] L.J. Chen, K.R. Fox, A.M. Haase, Body shape dissatisfaction and obesity among Taiwanese adolescents, Asia Pac. J. Clin. Nutr. 17 (2008) 457–460.
[21] S. Tucci, J. Peters, Media influences on body satisfaction in female students, Psicothema 20 (2008) 521–524.
[22] T.P. Gullota, G.R. Adams, C. Markstrom, The Adolescent Experience, fourth ed., Academic Press, San Diego, CA, 1999.
[23] D.W. Bunnell, P.J. Cooper, S. Hertz, I.R. Shenker, Body shape concerns among adolescents, Int. J. Eat. Disord. 11 (1992) 79–83.

[24] L.A. Proos, Anthropometry in adolescence -secular trends, adoption, ethnic and environmental differences, Horm. Res. 39 (1993) 18–24.
[25] D.C. Jones, T. Vigfusdottir, Y. Lee, Body image and the appearance culture among adolescent girls and boys: an examination of friend conversation, peer criticism, appearance magazines and the internalization of appearance ideals, J. Adolesc. Res. 19 (2004) 323.
[26] J.M. Connolly, V. Slaughter, L. Mealey, The development of preferences for specific body shapes, J. Sex Res. 41 (2004) 5–15.
[27] S.A. Sörensen, T.W. Teasdale, M.W. Higgins, The accuracy of reports of weight: children's recall of their parents' weights 15 years earlier, Int. J. Obes. 7 (1983) 115–122.
[28] I.D. Murdey, N. Cameron, S.J.H. Biddle, S.J. Marshall, T. Gorely, Pubertal development and sedentary behavior during adolescence. (Electronic version), Ann. Hum. Biol. 31 (2004) 75–86.
[29] T.M. Davison, Adolescent body image and psychosocial functioning, J. Soc. Psychol. 146 (2006) 15–30.
[30] A. Mussap, Waist-to-hip ratio and unhealthy body change in women, Sex Roles 56 (2007) 33–43.
[31] Y. Li, X. Hu, W. Ma, J. Wu, G. Ma, Body image perceptions among Chinese children and adolescents, Body Image 2 (2005) 91–103.
[32] M. Konstanski, A. Fisher, E. Gullone, Current conceptualisation of body image dissatisfaction:have we got it wrong? J. Child Psychol. Psychiatry 45 (2004) 1317–1325.
[33] M.P. McCabe, L.A. Ricciardelli, A longitudinal study of body image and strategies to lose weight and increase muscles among children, J. Appl. Dev. Psychol. 26 (2005) 559–577.
[34] K.L. LaBat, M.R. DeLong, Body cathexis and satisfaction with fit of apparel, Cloth. Text. Res. J. 8 (1990) 43–48.
[35] T. Tselepis, H.M. De Klerk, Early adolescent girls' expectations about the fit of clothes: a conceptual framework, J. Fam. Ecol. Consume. Sci. 32 (2004) 83–94.
[36] C.J. Salusso, Discussant: fit and sizing, in: C. Ladisch (Ed.), Proceedings of the International Textile and Apparel Association, 1995. Pasadena.
[37] D. Mellor, M. McCabe, L. Ricciardelli, J. Yeow, N. Daliza, N.F.M. Hapidzal, Sociocultural influences on body dissatisfaction and body change behaviors among Malaysian adolescents, Body Image 6 (2009) 121–128.
[38] L.J. Connell, P. Ulrich, K. Simmons, D. Pascoe, D. Bruner, Apparel Product Development for Plus-Sized Tween and Teen Boys, Project No. S06–AC03 (Competency: Management Systems), 2006.
[39] C. Hsu, Data mining to improve industrial standards and enhance production and marketing: an empirical study in apparel industry, Expert Syst. Appl. 36 (2009) 4185–4191.
[40] B. De Ville, Decision Trees for Business Intelligence and Data Mining: Using SAS North Carolina, SAS Enterprise Miner, 2006.
[41] M. Eckman, M.L. Damhorst, S.J. Kadolph, Toward a model of the in-store purchase decision process: consumer use of criteria for evaluating women's apparel, Cloth. Text. Res. J. 8 (1990) 13–22.
[42] ISO/TR10652:1991, Standard Sizing Systems for Clothes, 1991.
[43] EN 13402, Size Designation of Clothes, 1996.
[44] O. Kwona, K. Jungb, H. Youb, H.-E. Kimc, Determination of key dimensions for a glove sizing system by analyzing the relationships between hand dimensions, Appl. Ergon. 40 (2009) 762–766.
[45] A. Petrova, Creating Sizing Systems, Woodhead Publishing, 2007.
[46] ISO TC 133, The ISO Technical Committee on Sizing Systems and Designations for Clothes, 1991.

[47] M.F. Schilling, A.E. Watkins, W. Watkins, Is human height bimodal? Am. Stat. 56 (2002) 223–229.
[48] C.J. Wild, G.A.F. Seber, Chance Encounters: A First Course in Data Analysis and Inference, Wiley, New York, 2000.
[49] J. Devore, R. Peck, Statistics: The Exploration and Analysis of Data, Duxton, Belmont, 1997, p. 43.
[50] B. A'Hearn, P. Franco, V. Giovanni, Height and the normal distribution: evidence from Italian military data, Demography 46 (2009) 1–25.
[51] Business Wire, High Growth Forecasted in Global Children's Wear Market, 2007.
[52] R. Otieno, C. Fairhurst, The development of new clothing size charts for female Kenyan children. Part I: using anthropometric data to create size charts, J. Text. Inst. 1754–2340 91 (2000) 143–152.
[53] W.F.F. Kemsley, Reviewed Work(s): Women's Measurements and Sizes by Joint Clothing Council, Ltd. "The Incorporated Statistician", vol. 8, pp. 143–145 1958. http://www.jstor.org/pss/2986415.
[54] A. Petrova, Creating sizing systems, in: S.P. Ashdown (Ed.), Sizing in Clothing Developing Effective Sizing Systems for Ready-to-Wear Clothing, Woodhead Publishing, Cambridge, 2007, pp. 57–82.
[55] H. Hamid, KUB Tekstil to boost production by 16pc. Available at: www.kub.com.my/kubnews/news/NST20031129l.htm, In: NST.
[56] M.E. Faust, S. Carrier, A proposal for a new size label to assist consumers in finding well-fitting women's clothing, especially pants: an analysis of size USA female data and women's ready-to-wear pants for North American companies, Text. Res. J. 79 (16) (2009) 1446–1458.
[57] R. Otieno, C. Fairhurst, The development of new clothing size charts for female Kenyan children. Part II: size-range categorization and presentation of final size charts, J. Text. Instit. 91 (2000) 153–162.
[58] M.E. Faust, S. Carrier, P. Baptiste, Variations in Canadian women's ready-to-wear standard sizes, J. Fash. Market. Manag. 10 (2006) 71–83.
[59] M.J. Chung, M.J. Wang, The development of sizing systems for school students, in: In 36th International Conference on Computers and Industrial Engineering. Taipei, Taiwan, 2006.
[60] M. H, Q. He, J. Thornton, J. Wang, R.N. Pierson Jr., S. Heshka, D. Gallagher, Sex and race differences in fat distribution among Asian, African-American, and Caucasian prepubertal children, J. Clin. Endocrinol. Metab. 87 (2002) 2164–2170.
[61] E.V.D. Poel, A.R. Hosseinpoor, N. Sopeybroeck, T.V. Ourti, J. Vega, Socioeconomic inequality in malnutrition in developing countries, Bull. World Health Organ. 86 (2008) 241–320.
[62] K.M. Robinette, S.A. Mellian, C.A. Ervin, Development of Sizing Systems for Navy Women's Uniforms, Facility (Technical Report NCTRF 183), Navy Clothing and Textiles Research, Natick, MA, 1990.
[63] J.M. Winks, Clothing Sizes: International Standardization, 1997.
[64] S.R. Daniels, P.R. Khoury, J.A. Morrison, Utility of different measures of body fat distribution in children and adolescents, Am. J. Epidemiol. 152 (2000) 1179–1184.
[65] R. Otieno, C. Harrow, G. Lea-Greenwood, The unhappy shopper, a retail experience: exploring fashion, fit and affordability, Int. J. Retail Distrib. Manag. 33 (4) (2005) 298–309.
[66] F. Bellisle, Effects of diet on behaviour and cognition in children, Br. J. Nutr. 92 (2004) S227–S232.

[67] R. Otieno, Approaches in researching human measurement: MMU model of utilising anthropometric data to create size charts, EuroMed J. Bus. 3 (2008) 63–82.
[68] D.J. Gunnell, S.G. Davey, S. Frankel, M. Kempel, T.J. Peters, Socioeconomic and dietary influences on leg length and trunk length in childhood: a reanalysis of the Carnegie (Boyd Orr) survey of diet and health in prewar Britain (1937-1939), Perinat. Epidemiol. 12 (1998) 96–113.
[69] A. Pornwasin, Nectec develops national body chart, in: The Nation, 2009. Bangkok.
[70] C.J. Salusso, A Method for Classifying Adult Female Body Form Variation in Relation to the US Standard for Apparel Sizing, 1982.
[71] B.A. Yarnold, Principal Components Analysis and Exploratory and Confirmatory Factor Analysis, 1995.
[72] J. Robotham, Children's changing shape, in: Brisbane Times, 2009.
[73] I. Emanuel, M. Alexander, E. Churchill, B. Truett, A Height-Weight Sizing System for Flight Clothing, 1959.
[74] T.M. Plumlee, T.J. Little, Consumer purchase data as strategic product development tool, J. Text. Appar. Technol. Manag. 1 (2001) 1–10.
[75] M.J. Chung, H.F. Lin, J.J. Mao, J. Wang, The development of sizing systems for Taiwanese elementary-and high-school students, Int. J. Ind. Ergon. 37 (2007) 707–716.
[76] O. Barli, D. Elmali, R. Midilli, E. Aydintan, S. Ustün, A. Sagsöz, S. Ozgen, T. Gedik, Anthropometry of male and female children in Crèches in Turkey, Coll. Antropol. 29 (2005) 45–51.
[77] F. Ibrahim, B. Hinz, F. Seidel, F. Hofmann, G. Menzel, The significance of using anthropometric parameters and postures of European drivers as a database for finite-element models when calculating spinal forces during whole-body vibration exposure, Int. J. Ind. Ergon. 38 (2008) 816–843.
[78] D. Davidson, A.D.W. Thill, D. Lash, Male and female body shape preferences of young children in the United States, Mainland China and Turkey, Child Study J. 32 (2002).
[79] J.M. Lee, D. Appugliese, N. Kaciroti, R.F. Corwyn, R.H. Bradley, J.C. Lumeng, Weight status in young girls and the onset of puberty, Pediatrics 119 (2007) 624–630.
[80] K.M. Thompson, Changes in Children's exposure as a function of age and the relevance, Medscape Gen. Med. J. 6 (2004) 1–21.
[81] T. Baghurst, D.B. Hollander, B. Nardella, G.G. Haff, Change in sociocultural ideal male physique: an examination of past and present action figures, Body Image 3 (2006) 87–91.
[82] J.C. Yoon, C. Jasper, Key dimensions of women's ready-to-wear apparel: developing a consumer size-labeling system, Cloth. Text. Res. J. 14 (1996) 89–95.
[83] S. Ashdown, S. Loker, C. Andelson, Use of Body scan data to design sizing systems based on target markets, in: National Textile Center Research Briefs-management Systems Competency: NTC Project: S01-cr01, 2004, pp. 1–10.
[84] A. Beazley, Size and fit: formulation of body measurements tables and sizing systems, J. Fash. Market. Manag. 2(3) (1998) 260–284.
[85] R. Otieno, C. Fairhurst, The Development of New Clothing Size Charts for Female Kenyan Children. Part I: Using Anthropometric Data to Create Size Charts, J. Text. Instit. 91 (2) (2002) 1754–2340.

Selecting key body dimensions for the design of children's and teenagers' apparel

5

5.1 Introduction

This entire chapter presents the first step in sizing analysis process which is the key dimensions identification shown in Fig. 5.1. From Fig. 5.1, it can be seen that there are two steps involved for data analysis: multivariate analysis and factor analysis. The process for multivariate analysis involves validity test and percentile analysis. The validity test uses the KMO technique while the percentile analysis uses the percentage accommodation. After the validation of the data has been completed, the data is reduced using principal component analysis (PCA) method.

Section 5.2 presents the importance of the key dimensions for sizing system development. Section 5.3 describes the methodology and process of anthropometric data, which involves a few steps. The first step is to check the suitability of the data for multivariate analysis. The second step is to check the percentage accommodation of the data. Section 5.4 is the step to process the data for key dimensions identification selected from all variables of the body dimensions that describe the results of multivariate data examination. And lastly, Section 5.5 presents results of PCA. The goal of this chapter is to produce the selected key dimensions for different body parts and use these key dimensions to cluster the sample population according to the same groups' sizes (in the next chapter).

For each section, the tables of results reflect the four groups (male 7–12 years old and 13–17 years old; female 7–12 years old and 13–17 years old) following the segregation from Chapter 4.

5.2 The role of key body dimensions in apparel design

The anthropometric survey process was discussed in Chapter 3, which showed how it was planned and conducted. During the survey, there were many body dimensions taken from each person, including 50 different dimensions from top to toe (refer to Fig. 3.5). In order to group people according to their own sizes and shapes, we need to identify only a few significant body dimensions to divide the sample population into clusters having similar body dimensions. It is impossible to divide people according to 50 different dimensions. These body dimensions are known as the key dimensions. The objective of this analysis is to determine the key dimensions for different body

Clothing for Children and Teenagers. http://dx.doi.org/10.1016/B978-0-08-100226-1.00005-9
Copyright © 2016 Elsevier Ltd. All rights reserved.

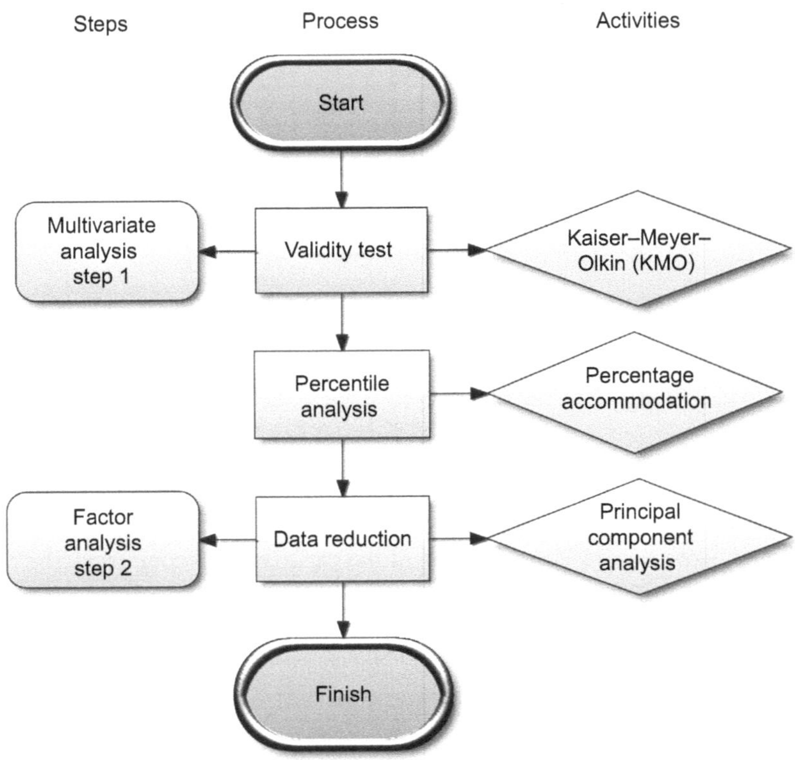

Figure 5.1 Process of key dimensions analysis.

parts so as to ensure the garment will fit well to the specific body parts such as upper body, lower body, and the whole body.

All the body dimensions collected from the sample population that resulted in a huge collection of data are statistically analyzed to identify the key dimensions [1]. Key dimensions can be different for different sample populations and for different types of garments. These are also used as control dimensions to assign the size of garment best suited for an individual for good fitting.

Control dimensions can be classified into primary control dimensions, secondary control dimensions, tertiary control dimensions, and so forth. Primary control dimensions are those that affect the goodness of fit in the garment and are the dimensions that are measured on a customer to match them with the right size garment. According to Winks [1], primary body dimensions are the body dimensions that are fundamental to body size [1]. These dimensions define the body size of a person and thus are the dimensions in which the sizing system is developed.

Secondary dimensions are those that are used together with the primary dimensions to define the body size of one person as a whole. For practicality, primary control dimensions need to be very familiar to the consumers; if consumers are familiar with the primary control dimensions, it will be easier for them to find their correct clothing size as the sizes are based on those key dimensions. To illustrate, for an upper body garment, if height is taken as the primary dimension, the population is first divided according to height and then further divided according to bust girth, if that is taken to be the secondary control dimension. After division, the size range of each population is obtained, which consists of the range of the two body dimensions used as key dimensions, range being defined as the upper and lower limit of the said dimensions. Researchers recommend the use of two or three control dimensions depending on garment type—the most common being the bust girth, waist girth, hip girth, and stature [2,3].

A combination of vertical and horizontal measurements must be used if the other measurements are to be predicted accurately [4]. Height seems most suitable as the predictor for vertical measurements. Weight, on the other hand, is the best predictor for girth measurements. Height and weight demonstrate the most variance within body length and thickness components, respectively. In other words, height and weight offer the best combination for accurately classifying adult female body size and shape [5]. However, researchers have consistently chosen bust girth (female) and chest girth (male) instead of weight for upper body and hip and waist for lower body [6]. Choosing these two dimensions for upper and lower body can provide better fit for garments [7].

5.3 Analysis of key body dimensions

All the 50 body dimensions taken from the samples (children ages 7–17) are tested for validity and reliability. All data were analyzed using the SPSS ver.13.0 statistical software. Bartlett's test of sphericity and the Kaiser–Meyer–Olkin (KMO) measure of sampling adequacy is performed to ensure the adequacy of sampling for PCA analysis. In this section, the methods to identify the suitability of the data for multivariate analysis are discussed. The percentile analysis is then conducted to understand the percentile value of the variables for better accommodation of the sample population. Then, the data will be analyzed using the multivariate data analysis technique. PCA is useful in assessing the dataset and allows understanding of the weight of the data or the aggregation.

5.3.1 Kaiser–Meyer–Olkin test

The initial test, namely the sampling adequacy test, was performed on the data prior to PCA. This is done to confirm the appropriateness of conducting PCAs in order to ensure that the data can be factored very well [8,9]. The KMO test is interpreted. The KMO statistic varies between 0 and 1. A value of 0 indicates that there is diffusion in the pattern of correlations, which shows that factor analysis is likely to be inappropriate. The value of 1 indicates that patterns of correlations are compact and thus factor analysis should yield distinct and reliable factors [10,11]. A KMO greater than 0.9 is considered to be marvelous [12,13].

In addition, Bartlett's test should reach a significance value to support the factorability of the correlation matrix obtained from the items [14]. Kaiser [15] claimed that the KMO statistic should be greater than 0.5. On the other hand, Hutcheson and Sofroniou [14] suggested that values between 0.5 and 0.7 are mediocre, values between

0.7 and 0.8 are good, values between 0.8 and 0.9 are great, and values above 0.9 are superb. In addition, for Bartlett's test, the value should be less than 0.05 in order for the test to be significant. If Bartlett's test showed value of less than 0.001, which is considered highly significant, it is then appropriate to conduct factor analysis.

5.3.1.1 Kaiser–Meyer–Olkin test results

From Table 5.1, the KMO results show all values are greater than 0.9 and Bartlett's test is highly significant since the observed significance level is 0.0000 ($p < 0.005$). For factor analysis, KMO is supposed to be greater than 0.5 and Bartlett test less than 0.05. As shown in Table 5.1, KMO values are 0.985 for male (7–12), 0.986 for female (7–12), 0.973 for male (13–17), and 0.965 for female (13–17). According to Raykov [16], KMO greater than 0.9 falls into the range of superb or marvelous. Thus, it can be concluded that the relationship between the variables is very strong and factor analysis can be carried out. In addition, the results also confirm that the data show sampling adequacy, which means it is likely to factor well for PCA analysis.

5.3.2 Percentile analysis

In the previous chapter, the extreme values were used to profile the samples, that is, the tallest, shortest, heaviest, and lightest. Analysis using extreme values often gives insight into the data and indeed the whole study. The percentile value, on the other hand, is the value below which a certain percentage of observations fall. It is used to ensure that the right percentage of sample population is taken for the development of the sizing system. The percentile division is first used on the group before the segmentation of sample population. In this study, the sizing system is designed according to the design limit, which accommodated 90% of the population.

The percentile between the 5th and 95th percentiles of the key dimensions—namely height, chest, waist, and hip girth—are shown in Table 5.2 for females and males ages 7–12. The range for the percentile value is more practical to build the sizing system on as compared to the extreme values. Having too wide of a range will result in a sizing system that is not that efficient because there will be some sizes that are accommodated by only a few people. After taking the percentile value, the range of the measurement is tolerable as compared with the extreme values, the range is too high.

Table 5.1 Sampling adequacy and reliability tests for all sample populations

Tests	Male (7–12)	Female (7–12)	Male (13–17)	Female (13–17)
KMO of sampling adequacy	0.99	0.99	0.97	0.97
Bartlett's test of sphericity	69,357.32	72,055.63	34,579.55	31,437.00
Approx. Chi-Square Df Sig.	1275	1326	1275	1378
($p < 0.005$)	0.000	0.000	0.000	0.000

Table 5.2 Percentile values of key dimensions according to gender, ages 7–12

	Male		Female	
7–12 years old	**5th**	**95th**	**5th**	**95th**
Height (cm)	114.2	151.5	113.8	154.0
Chest girth (cm)	56.3	90.1		
Bust girth (cm)			53.5	87.8
Waist girth (cm)	48.0	84.6	47.6	81.0
Hip girth (cm)	56.3	93.8	58.0	94.0

Table 5.3 Percentile values of key dimensions according to gender, ages 13–17

	Male		Female	
13–17 years old	**5th**	**95th**	**5th**	**95th**
Height (cm)	145.8	176.6	142.6	165.0
Chest girth (cm)	69.3	105.1		
Bust girth (cm)			69.0	98.5
Waist girth (cm)	56.5	94.5	55.7	86.0
Hip girth (cm)	72.6	106.8	76.1	108.4

Table 5.3 shows the percentile values of key dimensions for male and female (13–17 years old) samples. The 5th to 95th percentiles of height, bust, hip, and waist girth measurements for the sample size are tabulated and the range is reasonable for the development of size as compared to the extreme values.

5.4 Principal component analysis

PCA is the most common form of factor analysis and is categorized as a multivariate statistical technique. This method segments the variables into significant components. It reduced the variables for the selection of key dimensions, which are then used to segment the sample population. Important body variables were extracted with PCA method using Varimax rotation. To identify the number of components to be retained, three criteria were considered, namely latent root criterion, percentage of variance criterion, and Scree test criterion. The Eigenvalue >1 criterion was used. This method is used to analyze interrelationships among a large number of variables [17].

The objective of using PCA in this study was to reduce the number of variables and to cluster these variables into more parsimonious and manageable groups. Parsimonious means to summarize most of the original information (variance) in a minimum number of components for prediction purposes [15]. It also means that the variations of body dimensions are described using the fewest principal components (PCs) possible [10]. The body dimensions of each sample group are extracted using PCA and Varimax rotation. This is a common technique applied on anthropometric data to describe variations in the human body in a parsimonious manner by many previous sizing studies [2,18–23].

These groups are known as the components (factors). Each component contains interrelated variables [24]. Thus, from these components, the key dimensions can be selected. Subsequently, these key dimensions can be used to cluster the sample population into homogenous subgroups. The process flow of each step of PCA is described following and is shown graphically in Fig. 5.2.

5.4.1 The results of component analysis

In this section, how to extract the data using PCA method is discussed step by step using the real data collected from anthropometric survey. Since the data belong to the children, the data need to be separated into different age groups and gender, which is recommended from the findings in Chapter 4. The reason for this is because children have distinct different growth rates based on age and gender. In addition, the growth rates are really extensive as compared to adults who have steady growth rates; in fact, after the age of 20, adults do not have much physical growth.

Using the PCA method, all of the 50 anthropometric variables are input into the SPSS software in order to reduce them into components. In general, with 50 variables, you will extract 50 components in order to explain 100% of the variance in the data. The summarization of variables proves to be very good if the number of principal components is the same as the number of original variables. This indicates that the original information is not neglected as a variation factor [25].

According to Pedhauzer and Schmelkin [26], the first few components should extract at least 50% of the variance in order to prove the usefulness of PCA technique. The current study shows that 76% of the variance in the 50 body dimensions for females (ages 7–12) is explained in component one while for male samples (ages 7–12) this is 73%. As for the male and female samples between 13 and 17, lower percentage values are observed: 72.7% and 54%, respectively. Thus, all results indicate variance of more than 50% in the first component. This indicates that PCA technique is an effective method for this study to obtain a parsimonious solution in describing the variations of body shapes in children.

Furthermore, according to Hair et al. [17], the extracted factor should at least be 90% of the explained variance in order to show efficiency. From this finding, 90% variance was explained by eight principal components for male samples and seven principal components for female samples ages 7 to 12, respectively. In addition, for male and female samples (ages 13–17), 11 and 14 components, respectively, show 90% of variance. In contrast, a study conducted by Salusso et al. [27], only 60% of total variance for 15 components are found in the study, which is considered low. However, in 2006 the same researcher improved her studies by obtaining a 60% total variance accounting for two components [28].

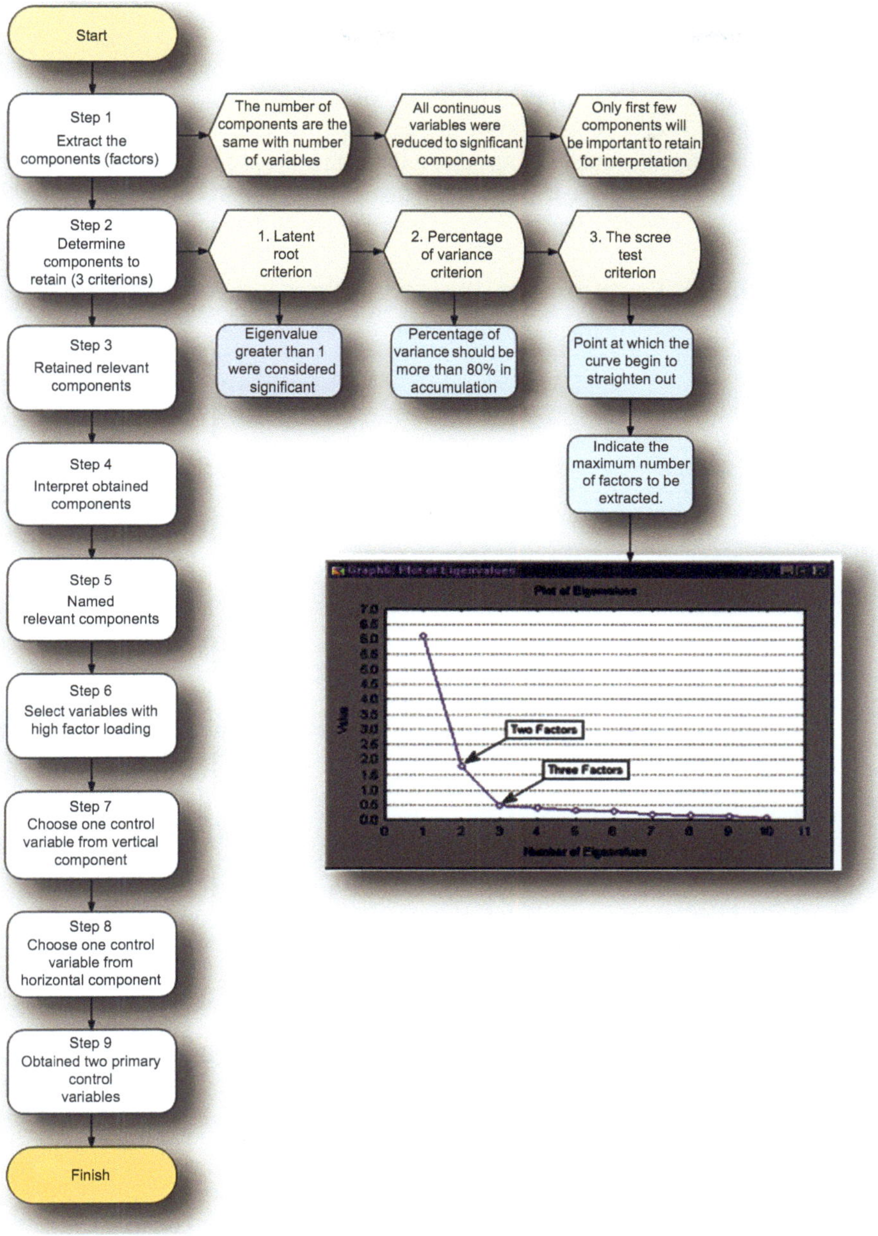

Figure 5.2 The principal component analysis flow chart.
Adapted from Ph.D. thesis of Norsaadah Zakaria.

In order to reduce the numbers of components for a more parsimonious solution, the criteria of retaining components are applied which are latent root, scree plot, and percentage of accumulated variance. All these criteria and results will be discussed in the following sections.

5.4.1.1 Latent root analysis

For this criterion, all the components extracted must have a value greater than 1 to be retained as the component for further analysis. Tables 5.4–5.7 show all the components that have been extracted from the anthropometric data with an eigenvalue greater than 1 (with bold values). Thus, this implies that these components are suitable to be retained. In all the tables, it can be seen that principal component 1(PC1) has the highest eigenvalue, which means that these are the PCs of the data set.

Table 5.4 Principal component analysis extraction for female (ages 7–12)

Component	Initial eigenvalues			Rotation sums of squared loadings	
	Total	% of variance	Cumulative (%)	% of variance	Cumulative %
1	**38.1**	76.3	76.3	41.9	41.9
2	**3.6**	7.2	83.5	77.5	77.5
3	**1.5**	2.9	86.4	8.4	**86.4**

Table 5.5 Principal component analysis extraction for male (ages 7–12)

Component	Initial eigenvalues			Rotation sums of squared loadings	
	Total	% of variance	Cumulative (%)	% of variance	Cumulative %
1	**36.3**	72.7	72.7	43.6	43.6
2	**4.5**	9.1	81.7	33.9	77.6
3	**1.2**	2.3	84.1	5.5	83.3
4	**1.0**	2.0	86.1	3.1	**86.1**

Table 5.6 Principal component analysis extraction for male (ages 13–17)

Component	Initial eigenvalues			Rotation sums of squared loadings	
	Total	% of variance	Cumulative (%)	% of variance	Cumulative %
1	**30.7**	61.4	61.4	40.9	40.9
2	**6.9**	13.6	75.0	24.3	65.3
3	**1.7**	3.5	78.5	11.4	76.7
4	**1.2**	2.4	80.9	3.8	80.5
5	**1.1**	2.1	83.0	2.6	**83.0**

Table 5.7 Principal component analysis extraction for female (ages 13–17)

Component	Initial eigenvalues			Rotation sums of squared loadings	
	Total	% of variance	Cumulative (%)	% of variance	Cumulative %
1	**27.0**	54.0	54.0	39.3	39.3
2	**7.5**	15.0	69.0	20.8	60.0
3	**2.0**	4.1	73.1	7.5	67.6
4	**1.9**	3.7	76.9	6.6	74.1
5	**1.1**	2.3	79.1	3.6	77.8
6	**1.0**	2.0	81.1	3.4	**81.1**

As shown in Table 5.4, for female (ages 7–12), three components (1, 2, and 3) show eigenvalue greater than one in which component 1 is the highest value of 38.1%, followed by 3.6% and 1.5%, respectively. The factor rotation value for retained components is observed at cumulative % of **86.4%** and **86.0%** for female and male, respectively. As for male samples in Table 5.5, four components are retained with eigenvalue of 36.3%, 4.5%, 1.2%, and 1.0%, respectively. With few components retained, the cumulative % show good total explained values.

For the male population ages 13–17 years old, the full extracted results with eigenvalue more than 1 are shown in Tables 5.6 and 5.7. From Table 5.6, the first five components (1, 2, 3, 4, and 5) are retained for an eigenvalue greater than 1. The first component shows eigenvalue of 30.7% followed by 6.9%, 1.7%, 1.2%, and 1.1%, respectively. The five rotation sums loading show cumulative % of **83.0%**.

Lastly, for the female population ages 13–17 years old, the result is shown in Table 5.7. In this table, six components (1, 2, 3, 4, 5, and 6) show an eigenvalue greater than 1. The first component shows eigenvalue of 27.0% followed by 7.5%, 2.0%, 1.9%, 1.1%, and 1.0% in which the total rotation sums of squared loadings account for **81.1%**.

5.4.1.2 Scree plot analysis

For criteria two, which is the scree plot, the results are shown in Figs. 5.3–5.6. An elbow where the curve merges into a straight line can be seen at components 2 and 3 (which are marked by red circles). The component at that merge location is the point to stop. From Fig. 5.1, the elbow is obvious from PC2 and PC3, and thus PC1 to PC3 are retained. Similarly, for all the other groups, three components are retained (Figs. 5.3–5.6).

5.4.1.3 Percentage of accumulated variance

The third criterion for retaining values depends on the percentage of accumulated variance. Based on the examination of eigenvalue and scree plot, three components

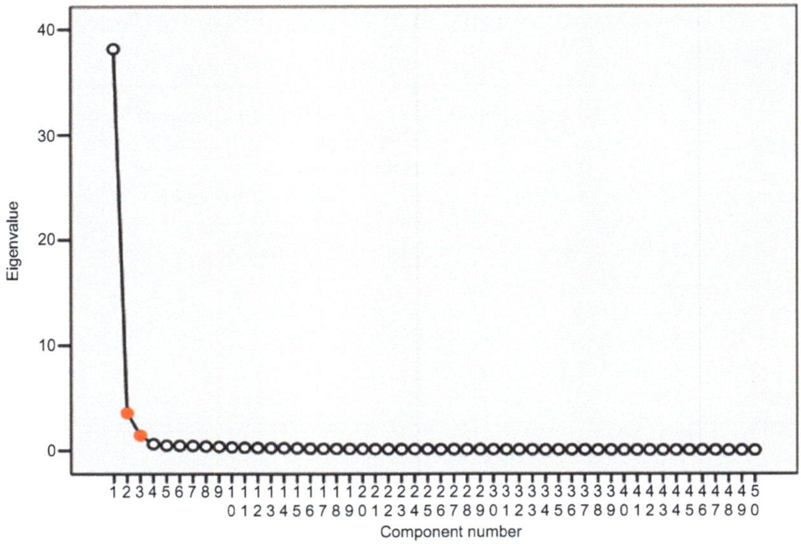

Figure 5.3 Scree plot of female (7–12 years old).

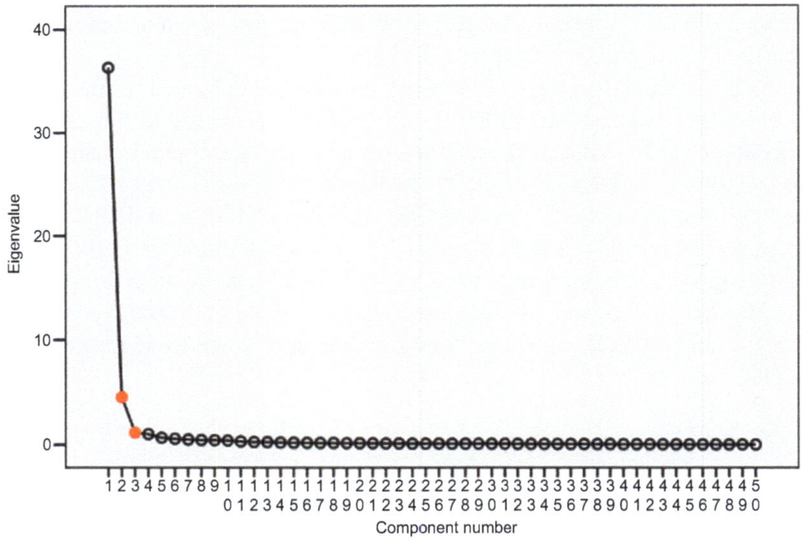

Figure 5.4 Scree plot of male (7–12 years old).

are retained for all samples groups with rotation sums of value of **86.4%** for females (7–12 years old); **83.3%** for males (7–12); **76.7%** for males (13–17), and **76.7%** for females (13–17). It can be observed that the cumulative percentage is reduced when fewer components are retained. However, at this point the numbers of components are not finalized yet until all the factor loadings of each component are examined, which clearly distinguish those variables that correlate highly with each component.

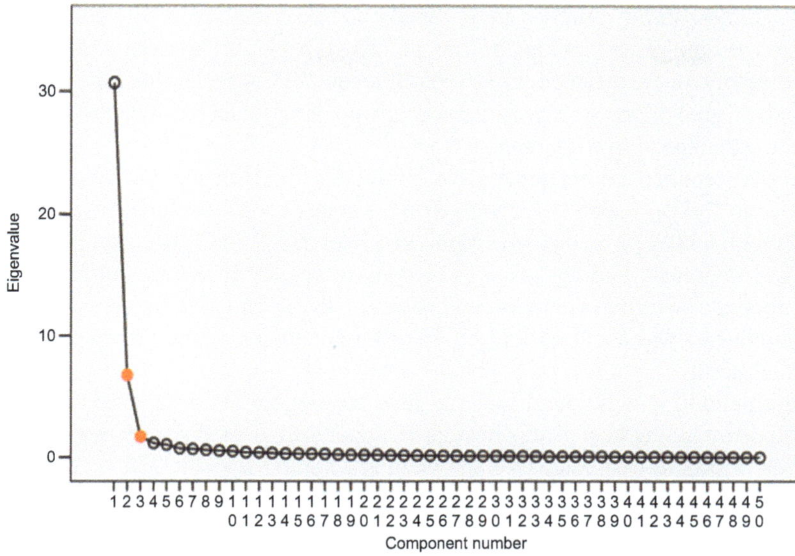

Figure 5.5 Scree plot of male (13–17 years old).

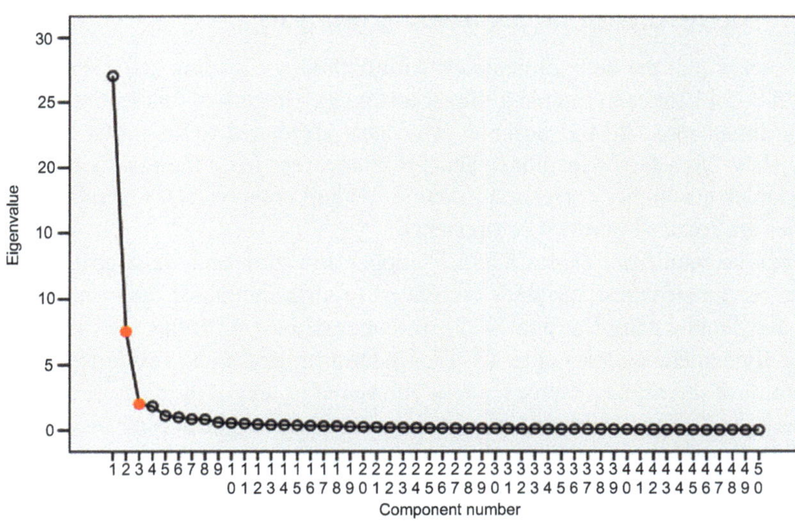

Figure 5.6 Scree plot of female (13–17 years old).

In general, previous studies have shown that body dimensions are accumulated into its own component which can be interpreted as one type of measurements such as the length, girth, or width. After analysis of factor loading, the current findings yield two significant components, which consist of variables of the highest factor loading. Thus, two components are retained namely the principal component 1 (PC1) and principal component 2 (PC2) for each sample population groups. For sample group of female (7–12), 46 variables out of 50 are clustered within two components while for male

(7–12) samples, 49/50 variables are grouped into two components. In addition, for the other two sample groups: male and female (13–17), 44/50 variables and 42/50 variables respectively are grouped into two components. PCA technique has successfully achieved the goal of getting a parsimonious group of variables, which extracted almost all of the variables in two components.

The two retained components gave cumulative percentages as follows: female (7–12) with **77.5%**; male (7–12) with **77.6%**; male (13–17) with **65.3%**; and female (13–17) with **60.0%**, which can be found in Tables 5.4–5.7 at rotated sums of squared loading. The present findings seem to be consistent with other research, which commonly retained two final components. However, the cumulative percentage for these two components was found differently in different studies ranging from 65% to 81% [2,20,21,23,29].

Consequently, it was found that the first component (PC1) consists of all girth dimensions including bust girth, chest girth, upper arm girth, hip girth, and waist girth and a few width dimensions like back width and shoulder width. The second component (PC2) consists of all length dimensions such as height, cervical height, upper arm length, and arm length. This finding is found to be similar to many previous sizing studies where two components represent the girth and length factors [2,20–22,29].

5.4.2 The results of factor loading analysis

In this section, all the body dimensions with high factor loading (≥0.75) are listed in Tables 5.8–5.11 for each sample group accordingly. The aim of this section is to select the key dimensions. All the variables are highly correlated to their own components if they show high factor loadings. Thus, it is apparent from the results that most of the variables are highly correlated to the individual components. Overall, almost all variables are reduced into two components.

As can be seen from Tables 5.8–5.11, upper arm girth and waist girth are distinguished as the strongest variables correlated to girth factor for male samples (ages 7–17) and female samples (age 7–12). In contrast, bust girth has the highest factor loading for female samples ages 13–17. Under arm length and inside leg length are noted to have the highest factor loading correlated to length in males ages 7–17 and females ages 7–12 as compared to arm length and hip height in females ages 13–17. Hence, PCA analysis for each sample group is shown following.

5.4.2.1 Male samples (ages 7–12)

In this sample group, referring to Table 5.7, there are 49 variables loaded on two components. However, in this section only significant variables with more than 0.75 factor loading are shown. It was found that for upper body, there are 13 variables and for lower body there are 14 variables.

Twenty-seven body dimensions show high factor loadings (≥0.75) in this study as compared to 11 dimensions (≥0.75) obtained from the Chung et al. study [18]. For this sample group, the highest factor loading is found in upper arm girth (0.91), waist girth (0.90), inside leg length (0.92), and hip height (0.9). In contrast, the highest factor

Table 5.8 Principal component analysis with varimax rotation for males (7–12 years old)

Body types	Variables	PC 1 (girth)	PC 2 (length)
	Height		**0.87**
	Weight	**0.86**	
Upper body (girth)			
	1. Upper arm girth	0.91	
	2. Elbow girth	0.88	
	3. Chest girth	0.87	
	4. Neck girth	0.87	
	5. Wrist girth	0.85	
	6. Armscye girth	0.84	
	7. Cervical to breast point	0.81	
	8. Neck base girth	0.79	
	9. Trunk circumference	0.81	
Lower body (girth)			
	10. Waist girth	0.90	
	11. Thigh girth	0.89	
	12. Calf girth	0.81	
	13. Hip girth	0.88	
	14. Mid-thigh girth	0.88	
	15. Lower knee girth	0.87	
	16. Knee girth	0.84	
	17. Crotch	0.85	
Upper body (length)			
	18. Under arm length		0.88
	19. Arm length		0.84
	20. Upper arm length		0.83
	21. Cervical to wrist		0.77
Lower body (length)			
	22. Inside leg length		0.92
	23. Hip height		0.90
	24. Thigh length		0.89
	25. Knee height		0.89
	26. Outer leg length		0.87
	27. Waist height		0.85
	Total proportion (%)	43.6	77.6

loading in the Chung et al. study is found only in chest girth (0.9) and chest girth below bust (0.85). In addition, only a few variables, namely height, chest girth, chest girth below bust, chest breadth, shoulder to elbow length, and biacromial breath have loading ≥0.8 was found in previous study as compared to 18 variables in this study [19]. It can also be observed from this study that more body dimensions with high

Table 5.9 **Principal component analysis with varimax rotation for females (7–12 years old)**

Body types	Variables	PC 1 (girth)	PC 2 (length)
	Height		**0.85**
	Weight	**0.83**	
Upper body (girth)			
	1. Upper arm girth	0.88	
	2. Back width	0.75	
	3. Bust girth	0.84	
	4. Armscye girth	0.85	
	5. Neck girth	0.83	
	6. Elbow girth	0.82	
	7. Wrist girth	0.78	
	8. Cervical to breast point	0.80	
	9. Neck shoulder to breast point	0.77	
	10. Trunk circumference	0.77	
Lower body (girth)			
	11. Waist girth	0.88	
	12. Mid-thigh girth	0.87	
	13. Thigh girth	0.86	
	14. Calf girth	0.86	
	15. Hip girth	0.83	
	16. Lower knee girth	0.84	
	17. Knee girth	0.79	
	18. Minimum leg girth	0.79	
	19. Crotch	0.85	
Upper body (length)			
	20. Hand length		0.79
	21. Upper arm length		0.86
	22. Cervical to wrist		0.76
	23. Under arm length		0.85
	24. Arm length		0.83
Lower body (length)			
	25. Knee height		0.86
	26. Inside leg length		0.90
	27. Hip height		0.88
	28. Waist height		0.83
	29. Outer leg length		0.82
	30. Thigh length		0.84
	Total proportion (%)	41.9	77.5

Table 5.10 Principal component analysis with varimax rotation for males (13–17 years old)

Body types	Variables	PC 1 (girth)	PC 2 (length)
	Height		**0.77**
	Weight	**0.91**	
Upper body (girth)			
	1. Upper arm girth	0.94	
	2. Neck girth	0.85	
	3. Elbow girth	0.87	
	4. Wrist girth	0.88	
	5. Chest girth	0.88	
	6. Armscye girth	0.87	
	7. Wrist girth	0.87	
	8. Neck shoulder to breast point	0.80	
Lower body (girth)			
	9. Waist girth	0.92	
	10. Thigh girth	0.93	
	11. Lower knee girth	0.94	
	12. Mid-thigh girth	0.93	
	13. Hip girth	0.91	
	14. Knee girth	0.91	
	15. Calf girth	0.92	
	16. Min leg girth	0.80	
	17. Ankle girth	0.76	
	18. Crotch	0.80	
Upper body (length)			
	19. Arm length		0.78
	20. Under arm length		0.82
	21. Upper arm length		0.81
Lower body (length)			
	22. Inside leg length		0.96
	23. Hip height		0.90
	24. Knee height		0.85
	25. Outer leg length		0.82
	26. Waist height		0.81
	27. Crotch		0.80
	28. Thigh length		0.80
	Total proportion (%)	40.9	65.3

factor loading are found, which implies that more variables can become the predictor of their components.

In fact, it was also observed that factor loadings of the height and weight of the samples is high in Table 5.8, which implies that body dimensions in this study are highly related to the components. High association between variable and component

Table 5.11 **Principal component analysis with varimax rotation for females (13–17 years old)**

Body types	Variables	PC 1 (girth)	PC 2 (length)
	Height		**0.77**
	Weight	**0.93**	
Upper body (girth)			
	1. Neck girth	0.81	
	2. Back width	0.81	
	3. Bust girth	0.93	
	4. Upper arm girth	0.93	
	5. Armscye girth	0.90	
	6. Elbow girth	0.85	
	7. Wrist girth	0.83	
Lower body (girth)			
	8. Waist girth	0.94	
	9. Hip girth	0.93	
	10. Thigh girth	0.92	
	11. Mid-thigh girth	0.91	
	12. Lower knee girth	0.91	
	13. Calf girth	0.88	
	14. Knee girth	0.84	
	15. Crotch	0.78	
Upper body (length)			
	16. Arm length		0.84
	17. Under arm length		0.82
	18. Cervical to breast point		0.8
	19. Neck shoulder to breast point		0.76
	20. Trunk circumference		0.77
Lower body (length)			
	21. Hip height		0.91
	22. Waist height		0.87
	23. Knee height		0.84
	24. Inside leg length		0.84
	25. Out leg length		0.83
	26. Thigh length		0.82
	Total proportion (%)	39.3	60.0

indicates that those variables can be the significant predictor of the component it belongs to. For example, an increase of waist girth can be associated with an increase in upper arm girth. Finally, this study proves efficiency of the two components at 77.6%. Higher total cumulative variance implies that the variables that belong to the components are more related to the components. It is also observed that this study obtained two components to describe the variations that exist in the body dimensions.

5.4.2.2 Female samples (ages 7–12)

For female samples, 46 variables are loaded on two components. Out of 46 variables, 30 variables are found to have high factor loadings (≥0.75). It was noted that there are 15 variables for upper body and another 15 variables for lower body, respectively.

5.4.2.3 Male samples (ages 13–17)

Tables 5.10 and 5.11 show significant variables for males and females ages 13–17 years old. In male and female samples, it was found that in total, 44 and 42 variables, respectively, out of 50 are loaded on two components.

5.4.2.4 Female samples (ages 13–17)

For male samples, 28 variables were found correlated to the girth and length component as compared to 26 variables in female samples. On the other hand, 18 and 15 variables are loaded on girths for males and females, respectively. Another 10 and 11 variables are loaded on lengths for males and females, respectively.

5.4.3 Identifying key dimensions

From the previous tables, PCA method confirms that those variables shown in Tables 5.8–5.11 prove to be the significant dimensions often taken in any anthropometric survey. The key dimensions commonly acknowledged in other literature are also found significant in this study, namely waist girth, bust girth, chest girth, and height. For example, bust girth has the highest factor loading for female samples ages 13–17. However, for children's data in this study, it is noted that upper arm girth is selected to be the strongest variable correlated to girth factor for male samples (ages 7–12 and 13–17) and female samples (ages 7–12), which has not previously been described in any sizing studies. Moreover, under arm length and inside leg length are noted to have the highest factor loading correlated to length in males ages 7–12 and 13–17 and females ages 7–12 as compared to arm length and hip height in females ages 13–17. Under arm length is also another variable found to have high relationship with length, which is not common to other sizing studies.

Overall, the girth variables chosen as the key dimensions are chest and bust girths for upper body. In contrast, hip girth is chosen for the lower body key dimensions. The selection of these body dimensions as the key dimensions confirms the findings from correlation analysis (from Chapter 4) in which the same common dimensions for girth are found to be significant as the key dimensions for upper body while for lower body the hip girth is chosen instead of waist girth. Hip was a better choice as compared to waist girth for some of these reasons; it was mentioned as a variable that has a variety of relationships with upper and lower torso [30,31]. Gordon [32] said that hip was a better selection for lower garments simply because this dimension cannot be easily adjusted after it has been made. Moreover, hip was also considered a better choice for the children as they are rapidly growing and in which alterations or adjustments are common during this age range [33]. From previous studies it has been noted that hip was a more stable measurement and found to have high correlation with girth component and therefore it is determined to be the key dimension for lower body [19,34].

In general, for all the tables (Tables 5.8–5.11), height is selected as the key dimension for both upper and lower body. Height is selected based on the high factor loading. The current study found that height represents almost 90% (0.87 and 0.85) of the length component for male and female samples in age group 7–12. On the other hand, for male and female samples ages 13–17 years old, the height represents almost 80% (0.77) of the length factors. This is consistent with the results from correlation analysis in Chapter 4 that show height is very strongly correlated with other length variables. From the opinion of other researchers, height is a must to be incorporated in a sizing system [35,36]. In addition, according to James and Stone [37], height was found an advantage especially for children as it could be easily measured in retail shops where a height chart could easily be obtained. Height is also regarded as a better estimator of size rather than age [38]. Hence, height is the most suitable dimension to cluster the samples into different groups.

All the key dimensions mentioned herein are finally selected for the classification of the sample population. These key dimensions are selected based on the high relationship with the main body measurements, girth and length. Furthermore, it has been mentioned in previous studies that key dimensions must be convenient to take [39]. This indicates that the selected key dimensions should be the ones that can be measured easily and practical when it comes to children [1]. In addition, it is also stated that if these measurements will be used as the size coding, the customers should be very familiar with the key dimensions [40]. For example, chest girth or bust girth for the upper body as compared to upper arm girth.

Upper arm girth is shown to have higher factor loading as compared to chest or bust girth but it is found to be unstable due to its small difference in measurements [40]. The key dimensions must also be easily recognized and familiar among customers [40,41]. The key dimensions are recommended to be an integral part of the garment [38]. Moreover, these are the common key dimensions often used by apparel experts and have proven to be useful and important in developing the new size tables based on scientific research [40–43].

As suggested by O'Brien and Shelton [41], stature should be combined with other measurements. Therefore, height and bust girth were identified as the key dimensions to segment female samples. In contrast, height and hip girth were chosen for the lower body measurement table for both genders and age groups. Moreover, these control variables are frequently used by the experts in clothing size charts (Table 5.12).

Table 5.12 Key dimensions for sample classification

Sample size	Key dimensions upper	Factor loading	Key dimensions lower	Factor loading
Male: 7–12	Height/chest girth	0.86/0.87	Height/hip girth	0.86/0.88
Female: 7–12	Height/bust girth	0.85/0.85	Height/hip girth	0.85/0.83
Male: 13–17	Height/chest girth	0.79/0.88	Height/hip girth	0.79/0.91
Female: 13–17	Height/bust girth	0.78/0.94	Height/hip girth	0.78/0.92

Table 5.13 **Body dimensions for both genders and age groups**

Upper body variables		Lower body variables	
Girth	*Length*	*Girth*	*Length*
Upper arm girth Armscye girth	Upper arm length Under arm length	Waist girth Mid-thigh girth Thigh girth Calf girth Hip girth Lower knee girth	Inside leg length Hip height Outer leg length Thigh length

5.5 Summary

The present results are significant in at least major two aspects. Firstly, PCA method was successfully used as a parsimonious summarization tool as almost all variables were extracted and clustered into only two components. It was also observed that almost 56% of the variables in this study are highly correlated (factor loading ≥0.75) to the principal component. Hence, it is reasonable to say that the majority of the body dimensions in children in this study are highly related to either girth or length factors.

Secondly, the most important highlights of Table 5.13 are the four common body measurements significantly describing both genders' upper body (7–17 years old): upper arm girth, armscye girth, under arm length, and upper arm length. These observations were not mentioned in the previous sizing study for children. Body dimensions at the area of garment sleeves were emphasized in this study. This is due to the fact that the factor loading of these body dimensions was found the highest in all sample groups. These dimensions gave an impact relatively between body and garment, which is known as the ease allowance. Good ease allowance will provide good garment fit.

Thus, this study suggests that special attention should be given to these dimensions as these are the most critical body parts to be considered when manufacturing garments for children in the age range of 7–17. Moreover, in research conducted by Ng and et al. [44], they found that dynamic allowance between garment and body are important to allow the wearer to perform certain movements. Without enough ease allowance movement can be restricted and thus create uncomfortable feelings especially in children.

On the other hand, another 10 body dimensions best describe the lower body of both males and females in these sample populations (7–17 years old), namely waist girth, mid-thigh girth, thigh girth, calf girth, hip girth, lower knee girth, inside leg length, hip height, outer leg length, and thigh length.

References

[1] J.M. Winks, Clothing Sizes: International Standardization, 1997.
[2] D. Gupta, B.R. Gangadhar, A statistical model for developing body size charts for garments, Int. J. Clothing Sci. Technol. 16 (2004) 458–469.

[3] J.E. Croney, An anthropometric study of young fashion students including factor analysis of body measurements, Man 12 (1977) 448–496.
[4] J. Chun-Yoon, C.R. Jasper, Key dimensions of women's ready-to-wear apparel: developing a consumer size-labeling system, Clothing Text. Res. J. 14 (1) (1996) 89–95.
[5] Y.S. Lee, K. Robinette, D. Veitch, Comparing samples from different countries for key sizing dimensions, in: Banff Wear Conference, 2007.
[6] C.J. Salusso-Deonier, M.R. DeLong, F.B. Martin, K.R. Krohn, A multivariate method of classify in body form variation for sizing women's apparel, Clothing Text. Res. J. 4 (1985) 38–45.
[7] S.P. Ashdown, An investigation of the structure of sizing systems, Int. J. Clothing Sci. Technol. 10 (5) (1998) 324–341.
[8] B.G. Tabachnick, L.S. Fidell, Using Multivariate Statistics, sixth ed., Pearson Education, Limited, 2012.
[9] B.G. Tabachnick, L.S. Fidell, Using Multivariate Statistics, fifth ed., Allyn and Bacon, Boston, 2007.
[10] A. Field, Factor analysis on SPSS, in: C8057(Research Methods II): Factor Analysis on SPSS, 2005, pp. 1–8.
[11] B.P. O'Connor, SPSS and SAS programs for determining the number of components using parallel analysis and Velicer's MAP test, Behav. Res. Methods Instrum. Comput. 32 (2000) 396–402.
[12] C.E. Lance, M.B. Marcus, L.M. Michels, The sources of four commonly reported cutoff criteria: what did they really say? Organ. Res. Methods 9 (2006) 202–220.
[13] M.A. Pett, N.R. Lackey, J.J. Sullivan, Making Sense of Factor Analysis: The Use of Factor Analysis for Instrument Development in Health Care Research, SAGE, California, 2003.
[14] Hutcheson, N. Sofroniou, The Multivariate Social Scientist: Introductory Statistics Using Generalized Linear Models, 1999.
[15] J. Pallant, SPSS Survival Manual, 2001.
[16] T. Raykov, G.A. Marcoulides, An Introduction to Applied Multivariate Analysis, 2008.
[17] J.F. Hair, R.E. Anderson, R.L. Tatham, W.C. Black, Multivariate Data Analysis, fourth ed., 1998.
[18] M.J. Chung, H.F. Lin, J.J. Mao, J. Wang, The development of sizing systems for Taiwanese elementary-and high-school students, Int. J. Ind. Ergonomics 37 (2007) 707–716.
[19] Z.B.R. Azouz, M. Shu, R. Lepage, Characterizing human shape variation using 3D anthropometric data, Visual Comput. 22 (2006) 302–314.
[20] H.C. Hsu, H.F. Lin, M.J. Wang, Development female size charts for facilitating garment production by using data mining, J. Chin. Inst. Ind. Eng. 24 (2006).
[21] C.H. Hsu, H.F. Lin, M.J. Wang, Developing female size charts for facilitating garment production using data mining, J. Chin. Inst. Ind. Eng. 24 (2007) 245–251.
[22] C. Hsu, Data mining to improve industrial standards and enhance production and marketing: an empirical study in apparel industry, Expert Syst. Appl. 36 (2009) 4185–4191.
[23] C.J. Deonier, M.R. DeLong, B. Frank, K.R. Krohn, A multivariate method of classify in body form variation for sizing women's apparel, Clothing Text. Res. J. 4 (1985) 38.
[24] B.A. Yarnold, Principal Components Analysis and Exploratory and Confirmatory Factor Analysis, 1995.
[25] G. Hutcheson, N. Sofroniou, Introductory Statistics Using Generalized Linear Models, Sage Publications, Thousand Oaks, CA, 1999.
[26] E.J. Pedhazer, L.P. Schmelkin, Measurement, Design, and Analysis: An Integrated Approach, Lawrence Erlbaum Assoc., Hillside, NJ, 1991.

[27] C.J. Salusso-Deonier, M.R. DeLong, A multivariate method of classify in body form variation for sizing women's apparel, Clothing Text. Res. J. 4 (1982) 38–45.
[28] C.J. Salusso, J.J. Borkowski, N. Reich, E. Goldsberry, An alternative approach to sizing apparel for women 55 and older, Clothing Text. Res. J. 24 (2006) 96–109.
[29] C.H. Hsu, M.J.J. Wang, Using innovative technology to establish sizing systems, Int. J. Innovation Learn. 2 (2005) 233–245.
[30] C.C. Gordon, Anthropometric sizing and fit testing of a single battle dress uniform for US Army men and women, in: Performance of Protective, 1986. Philadelphia, PA.
[31] A. Beazley, Size and fit. Formulation of body measurements tables and sizing systems, J. Fash. Mark. Manage. 2 (1998).
[32] R. Zheng, W. Yu, J. Fan, Development of a new chinese bra sizing system based on breast anthropometric measurements, Int. J. Ind. Ergonomics 37 (2007) 697–705.
[33] C.H. Hsu, Applying a bust-to-waist ratio approach to develop body measurement charts for improving female clothing manufacture, J. Chin. Inst. Ind. Eng. 25 (2008) 215–222.
[34] H.F. Lin, C.H. Hsu, M.J. Wang, Y.C. Lin, An application of data mining technique in developing sizing system for army soldiers in Taiwan, World Sci. Eng. Acad. Soc. (WSEAS) Trans. Comput. 7 (2008) 245–252.
[35] Y. Kang, H.S. Choi, W.H. Do, A study of the apparel sizing of children's wears. An analysis of the size increments utilized in children's wear based on an anthropometric survey, J. Korean Home Econ. Assoc. Engl. ed. 2 (2001).
[36] W.F.F. Kemsley, Women's Measurements and Sizes, HMSO London, 1957.
[37] R. James, P. Stone, Children's Wear Sizing Survey, Clothing and Allied Products Industrial Training Board, Leeds, UK, 1984.
[38] R. Otieno, C. Fairhurst, The development of new clothing size charts for female Kenyan children. Part I: using anthropometric data to create size charts, J. Text. Inst. 1754–2340 91 (2000) 143–152.
[39] A. Petrova, Creating sizing systems, in: S.P. Ashdown (Ed.), Sizing in Clothing Developing Effective Sizing Systems for Ready-to-wear Clothing, Woodhead Publishing, Cambridge, 2007, pp. 57–82.
[40] R. O'Brien, W.C. Shelton, Body measurements of American boys and girls for garment and pattern construction, in: US Dept. of Agriculture Miscellaneous Publication 366, M.P. 454, US Department of Agriculture, Washington, DC, 1941, p. 141.
[41] J. Fan, W. Yu, L. Hunter, Clothing Appearance and Fit: Science and Technology, Woodhead Publishing Limited, Cambridge, 2004.
[42] A. Beazley, Size and fit. Procedures in undertaking a survey of body measurements, J. Fash. Mark. Manage. 2 (1997) 55–85.
[43] R. Otieno, C. Fairhurst, The development of new clothing size charts for female Kenyan children. Part II: size-range Categorization and Presentation of final size charts, J. Text. Inst. 91 (2000) 153–162.
[44] R. Ng, S.P. Ashdown, A. Chan, Intelligent size table generation, in: Proceedings of the Asian Textile Conference (ATC), 9th Asian Textile Conference, Taiwan, 2007.

Clustering children and teenagers according to apparel size

6

6.1 Introduction

This chapter presents the third step of sizing development process. The process of the data is based on the flowchart 6.1 below. As can be seen from Fig. 6.1, there are two processes involved in this chapter. The first process is to identify the cluster group using the k-means technique. Then the cluster groups are verified using the percentage accommodation technique. Section 6.2 presents the clustering technique to segment the sample population into similar key dimensions. The k-means centroid method was used to obtain the different groups for each age group and gender. Section 6.3 describes the process to verify the cluster groups by looking at the percentage accommodation. Section 6.4 is the step to process the data for key dimensions identification selected from all variables of the body dimensions that describe the results of multivariate data examination, and lastly, Section 6.5 presents results of principal component analysis (PCA). The goal of this chapter is to produce the numbers of groups that

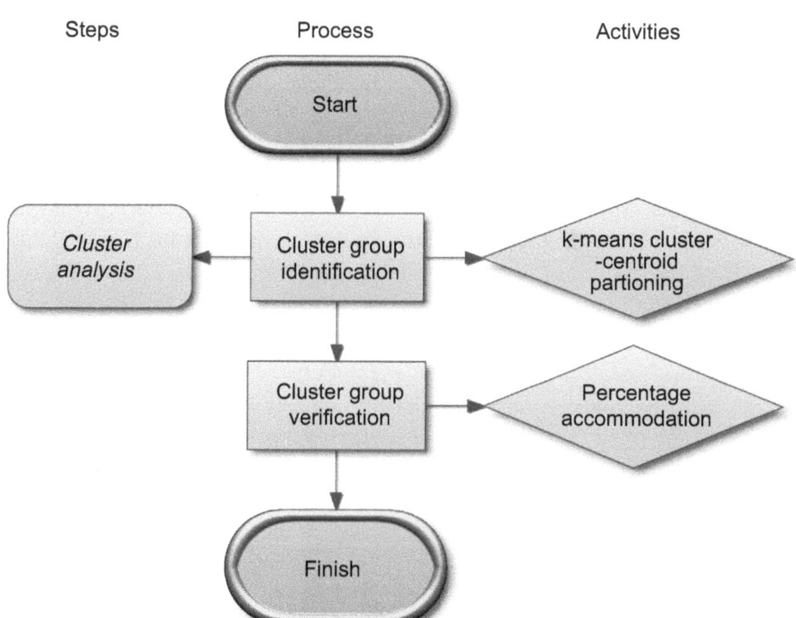

Figure 6.1 Process flowchart cluster analysis.

Clothing for Children and Teenagers. http://dx.doi.org/10.1016/B978-0-08-100226-1.00006-0
Copyright © 2016 Elsevier Ltd. All rights reserved.

have similar key dimension characteristics where all of the sample population will be grouped within their own body sizes and shapes.

6.2 The importance of distinct apparel sizes for children and teenagers

Cluster analysis is an exploratory data analysis tool used to segment people into homogenous subgroups. This means that each person in the group has similar physical traits in which people in one group differ from people in another group. The importance of clustering the sample population clarifies that there is a need to segment the sample population as children have different growth rates based on physical body dimensions and gender. Furthermore, based on the research, children do not grow the same although they are may be the same age. The growth of the children is very rapid and not stable. Therefore, they need to be segregated according to their sizes based on these differences. It is important to see the different cluster group that belongs to different age groups and to only classify them into the same traits of body dimensions.

6.3 Cluster analysis technique

The cluster analysis method used in this study is the k-means cluster; this method is chosen because it is regarded as a simple clustering method and often shows optimal results [1]. It is also proven to be a good technique in dividing the population into homogenous groups for identification of distinct body characteristics [1–3].

K-means clustering is a simple and easy way to classify a given data set through a certain number of clusters (assume k clusters) fixed a priori. First, the data set is divided into the number of clusters chosen (k), for example, if k is taken as 3, the data set is divided into three groups. Next, the data points are randomly assigned to the clusters. Suppose the data set has 30 data points; when it is partitioned into three groups, each cluster will have about 10 data points. For each data point, the distance from the centroid of the cluster is calculated. A data point that is far from its cluster is moved into the closest cluster. A data point that is close to its cluster remains where it is. This step is repeated until no data point is far from a cluster.

Although the procedure will always terminate, the k-means algorithm does not necessarily find the most optimal configuration corresponding to the global objective function minimum [4]. The algorithm is also highly sensitive to the initial randomly selected cluster centers. The k-means algorithm can be run multiple times to reduce this effect. The advantage of k-means clustering is that it can be easily adapted to available data by modifying feature attributes [4]. Clustering then may be processed in a multidimensional feature space. In the proposed algorithm, only geometric attributes are used to separate ground from off-ground points.

The movement of the centroid data points to the closest cluster only stops when every data point is closest to its own cluster. In the end, the data point is considered stable and the process of clustering has stopped. The k-means clustering process is illustrated in Fig. 6.2.

6.3.1 The identification of k factor clustering process

For this study, the k is determined to be three. The three clusters are chosen to reflect the typical sizes of clothing: small, medium, and large. Three cluster groups also have been proven to show excellent results to distinguish the characteristics from one another in previous studies [5–7]. The classification of people in each different group should ensure that in each group the members are similar based on the variables

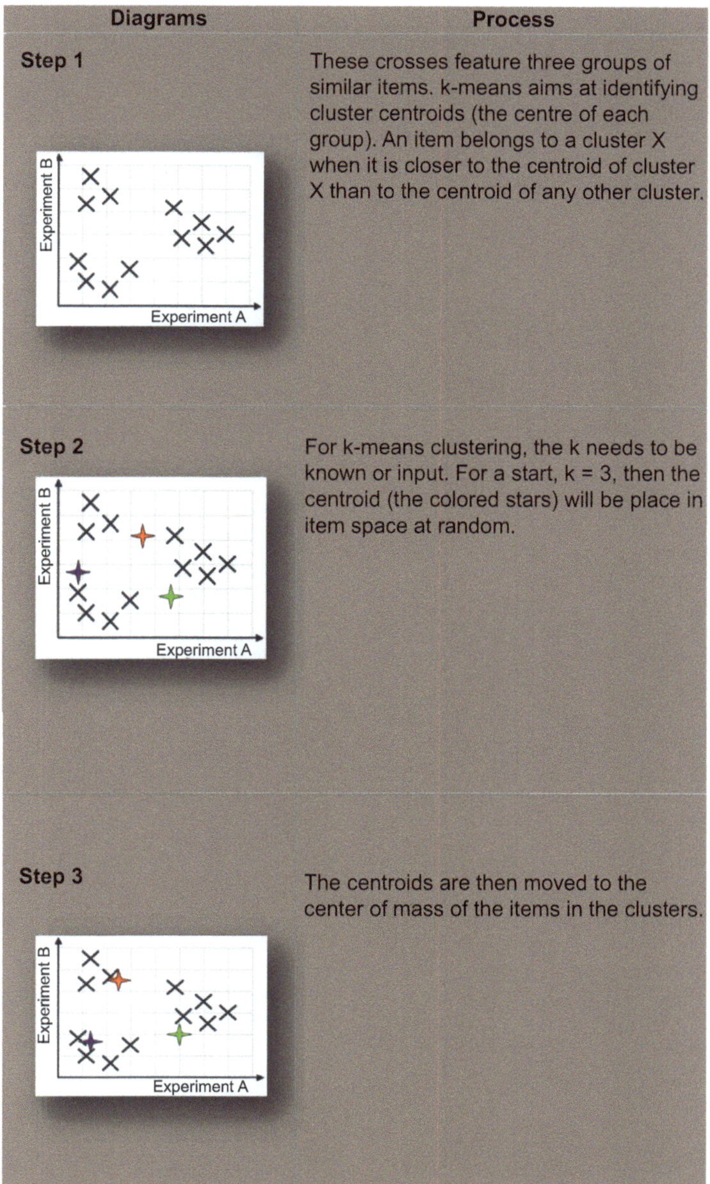

Figure 6.2 K-means clustering process.

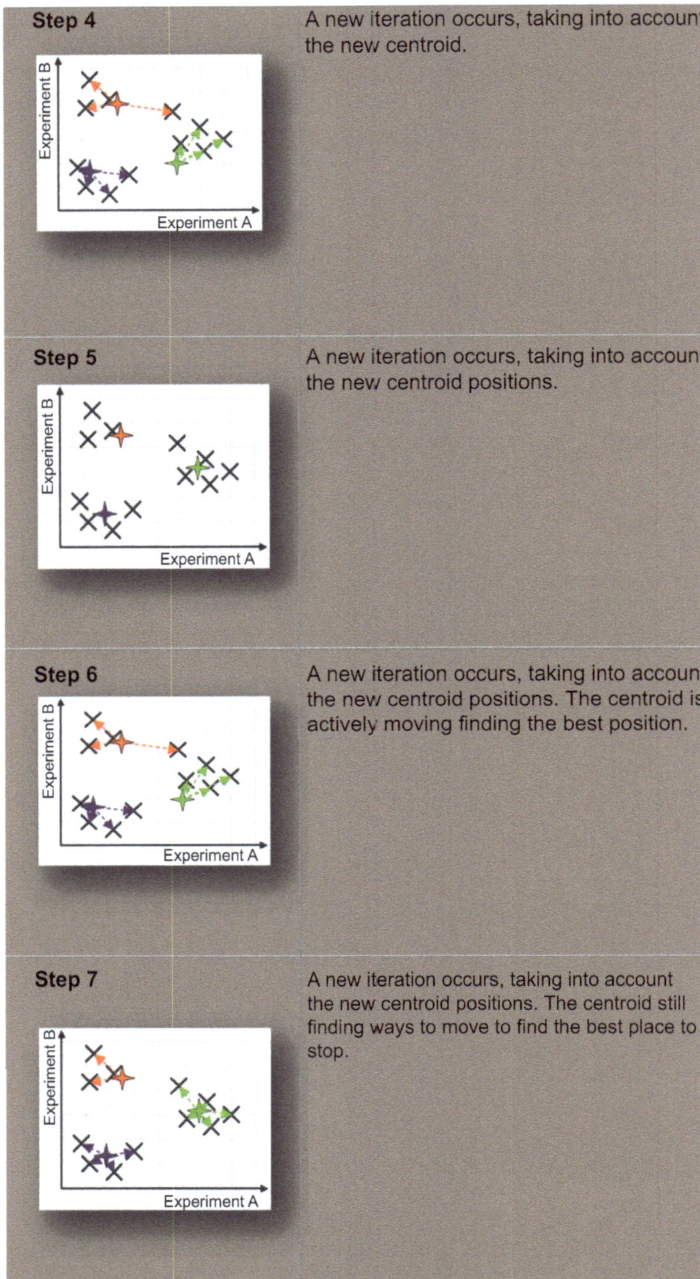

Figure 6.2 Continued

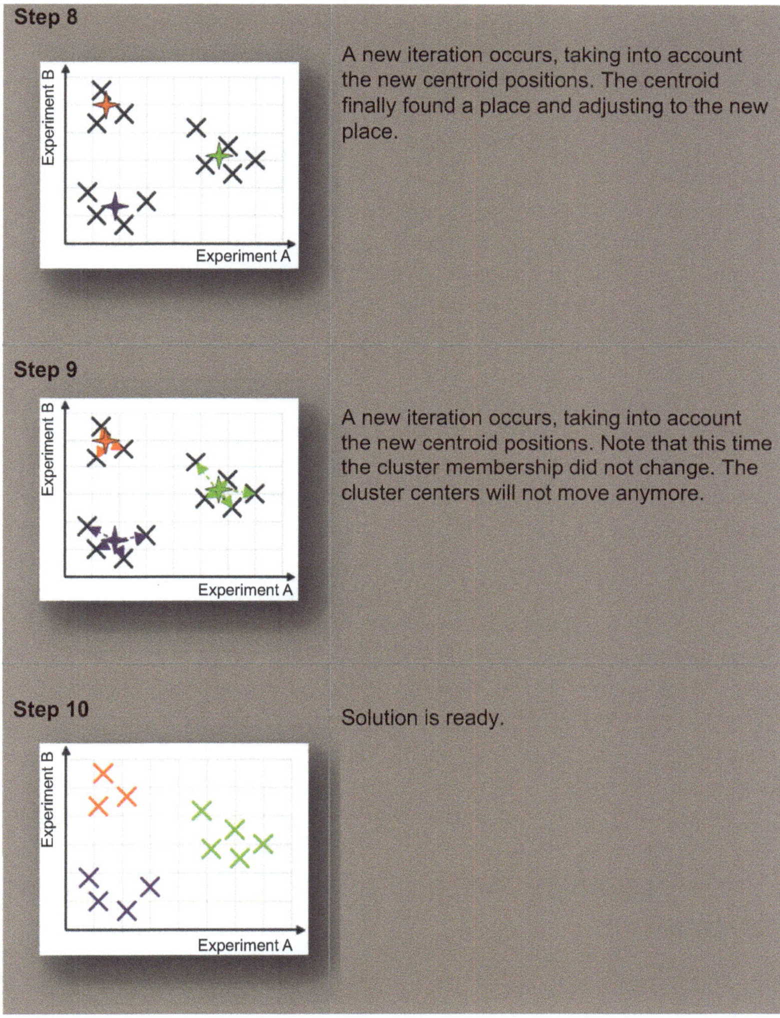

Figure 6.2 Continued

chosen to segregate them. For example, for the population of females ages 7–12 years old, the k is decided for three groups. The control variables are height and bust girth, so females who have similar height and bust measurements are classified into the same group. The cluster analysis technique stops the classification before more variables are found in the clustered groups.

The only drawback of using this type of cluster analysis is that although a k value must be chosen, there is no "right" number of k. The verification of the k number for the cluster groups was done using the data mining technique, as explained in the next section. In order to segment the children using k-means cluster, a certain number can be decided. The typical number of k can be decided by the researcher depending on the common sizes in their country or if there is a specific size pattern that can be seen from the analysis of body sizes and shapes.

Partitioning type of clustering (k-means clusters) was employed to analyze the anthropometric data for different gender and age groups (males and females, ages 7–12 and 13–17). From the PCA results, the selected key dimensions (height and girth variables) are used as the primary control dimensions, which means both height and girth dimensions are used to divide the sample population; this differs from many previous studies that used height as the primary control dimension followed by secondary control dimensions of bust, chest, or hip girth [8–10].

In order to create a garment that fits the body well, apparel experts recommend that size development should be designed according to the specific part of the body [11,12]. In addition, Gupta [13] adds that "better fit could be obtained in the garments if the sample is divided separately for the upper and the lower body." Thus, the clustering of the samples was divided into upper body and lower body. In Chapter 5, PCA demonstrated that height and girth are two important components. Therefore, height and bust girth or height and chest girth were used as the primary control dimensions for upper body classification, and height and hip girth were used for lower body classification.

In general, the k-means partitioning technique confirms that the grouping of the data into three different clusters is significantly different for each sample group. Similar to previous studies, cluster analysis technique was successfully used to divide the anthropometric data into three body types, which resulted in distinct groups [5,7]. In addition, the scatter plot of key dimensions for each sample group also depicts a clear, densely packed distribution of the data separated into well-defined body types. Three distinct body types were observed in each group of the sample population: small size, medium size, and large size.

6.4 The outcome of clustering in a sample population of children

The resulting analysis is presented in the following discussion. The results are divided into the segmented groups using the clustering technique. For each group, for example, for males 7–12 years old, all the data are divided into three groups and the results are analyzed accordingly.

6.4.1 Male Sample Size (7–12 years old)

6.4.1.1 Clustering for upper body

Three groups according to height and chest girth are depicted in Table 6.1. The majority of the sample population ($N=234$) fall under Cluster 2, which is the small size. The second highest distribution of sample size ($N=197$) is in Cluster 1, the medium size. The last group ($N=132$) belongs to Cluster 3, which is the large size.

In addition, Fig. 6.3 shows a scatter graph that is distributed into three well-defined clusters. Clusters 1 and 3 have about the same height range (average to tall) but distinct chest measurements. Cluster 1 show that samples are grouped under

Table 6.1 **Cluster groups for upper body: males (ages 7–12)**

	Cluster 2	Cluster 1	Cluster 3
N	234	197	132
Mean height (cm)	122.77	137.20	139.82
Range (cm)	114.5–132.7	127.0–151.0	121.0–151.2
Mean chest girth (cm)	61.52	67.22	80.64
Range (cm)	55.0–80.4	58.6–74.6	72.7–89.3
Body type	*Small*	*Medium*	*Large*

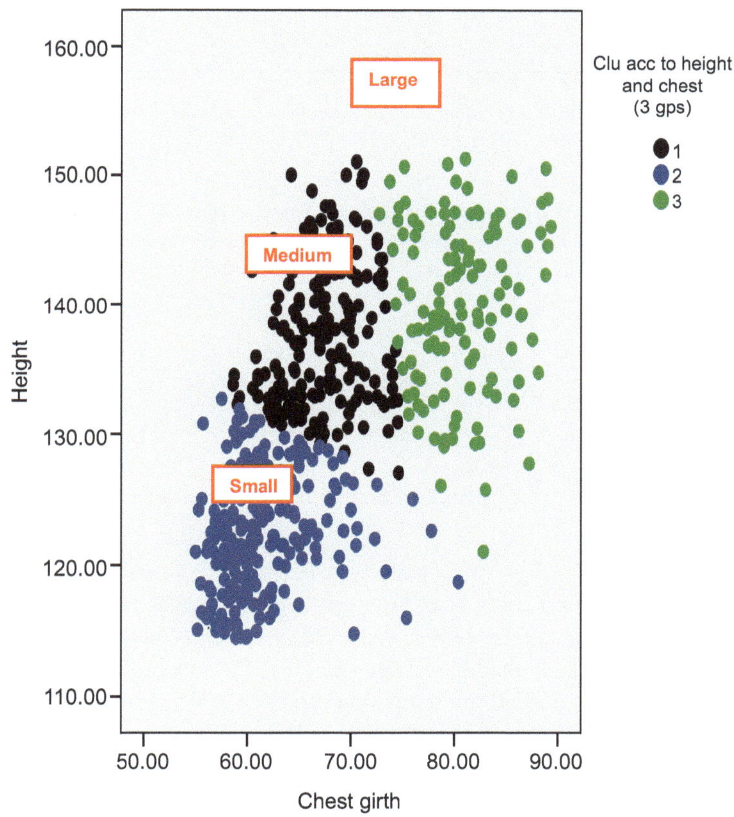

Figure 6.3 Scatter plot of cluster groups according to height and chest for males ages 7–12.

medium body type where they are average to tall with smaller chest measurements. Cluster 3, known as large size, is characterized by average to tall height and bigger chest. In Cluster 2, subjects are mainly found in the short height with small chest to big chest category.

Next, the samples were analyzed according to age categories. Table 6.2 shows the distribution of each age within the three cluster groups and shows that all ages are found in each cluster. However, for medium body type (Cluster 1) a majority of the

Table 6.2 **Age versus cluster groups for upper body: males (ages 7–12)**

Age	Cluster according to height and chest (three groups)			Total
	Cluster 2 (small)	*Cluster 1 (medium)*	*Cluster 3 (large)*	
7	87*	1	5	93
8	57	17	18	92
9	48	32	21	101
10	27	43	32*	102
11	11	48	32*	91
12	4	56*	24	84
Total	234	197	132	563

Note: For this and all similar tables in this section, * indicates the most frequently occurring age(s) in a given cluster. For example, the most frequent age in Cluster 1 is 12 ($N = 56$).

Table 6.3 **Cluster groups for lower body: males (ages 7–12)**

	Cluster 2	Cluster 3	Cluster 1
N	187	193	164
Mean height (cm)	122.2	133.6	141.7
Range (cm)	114.5–133.5	121.0–146.3	127.7–151.2
Mean hip girth (cm)	62.7	69.9	81.4
Range (cm)	56.3–77.4	60.1–85.5	70.0–93.3
Body type	*Small*	*Medium*	*Large*

samples belong to age 12. For small body type (Cluster 2), a majority of the samples are 7 years old. In Cluster 3, large size, a majority belong to ages 10 and 11. This result indicates that the lowest age samples (7 years old) are mainly in the small size group, which is expected; however, the highest age samples (12 years old) are not clustered in the large size group but instead in the medium body type. Those who belong to large size are mostly the middle age range, ages 10 and 11.

6.4.1.2 Clustering for lower body

Table 6.3 depicts the profile of each cluster group as follows: Cluster 1 with 164 samples, Cluster 2 with 187 samples, and Cluster 3 with 193 samples. Generally, the majority of the sample size falls under Cluster 3, medium size. The second highest distribution of sample size is in Cluster 2, which is the small size. The last group, Cluster 1, is the large size.

In addition, Fig. 6.4 gives a visual representation of this sample size showing three well-defined clusters based on height and hip girth. The results of this indicate that large size males are mainly tall with wide hips. Small size males are short with narrow to medium hips. Medium size males are found to be in the average to tall height range with medium-to-wide hips.

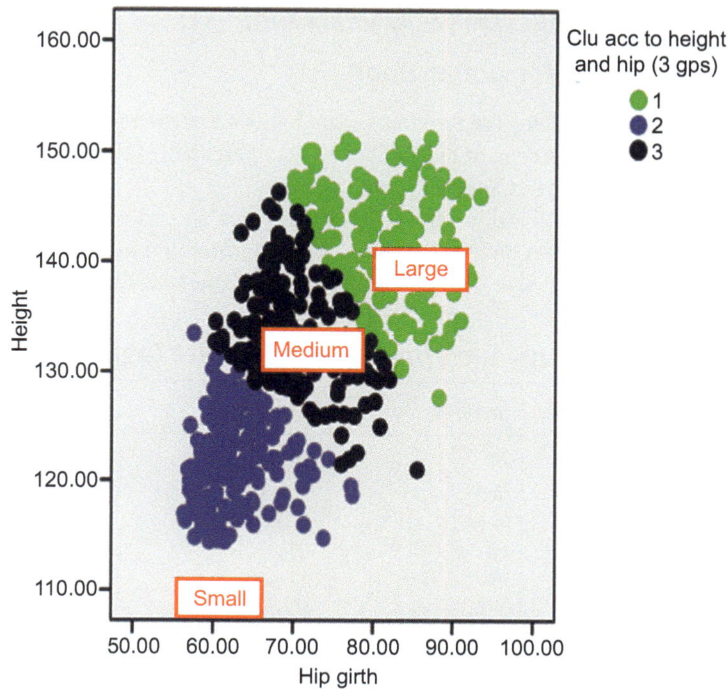

Figure 6.4 Scatter plot of cluster groups according to height and hips for males ages 7–12.

Table 6.4 Age versus cluster groups for lower body: males (ages 7–12)

Age	Cluster according to height and chest (three groups)			Total
	Small	*Medium*	*Large*	
7	*74	15	0	89
8	45	27	13	85
9	35	*40	24	99
10	21	39	39	99
11	8	*41	39	88
12	4	31	*49	84
Total	187	193	164	544

Note: For this and all similar tables in this section, * indicates the most frequently occurring age(s) in a given cluster. For example, the most frequent age in Cluster 1 is 12 (N=56).

From Table 6.4, the dominant age in each cluster group is as follows: the majority of the sample size falling under Cluster 1 (large size) are 12 years old; the majority of the samples in Cluster 2 (small size) are 7 years old; and the majority of the samples in Cluster 3 (medium size) are 9 and 11 years old.

6.4.2 Female sample size (7–12 years old)

6.4.2.1 Clustering for upper body

The centroids of the clusters for females (ages 7–12) are depicted in Table 6.5. The majority of the subjects are scattered in Cluster 1 (small size), followed by Cluster 3 (medium), and the fewest in Cluster 2 (large size).

Fig. 6.5 illustrates the distribution of female samples according to height (y-axis) and bust girth (x-axis). The well-defined clusters are separated into three main body types. Cluster 1, the small size group, has short height and small bust measurements.

Table 6.5 **Cluster groups for upper body: females (ages 7–12)**

	Cluster 1	Cluster 3	Cluster 2
N	197	194	134
Mean height (cm)	122.5	133.4	145.9
Range (cm)	114.0–131.0	120.5–146.0	132.1–153.7
Mean hip girth (cm)	59.4	67.6	77.1
Range (cm)	53.5–71.8 cm	57.3–83.0	63.0–87.5
Body type	*Small*	*Medium*	*Large*

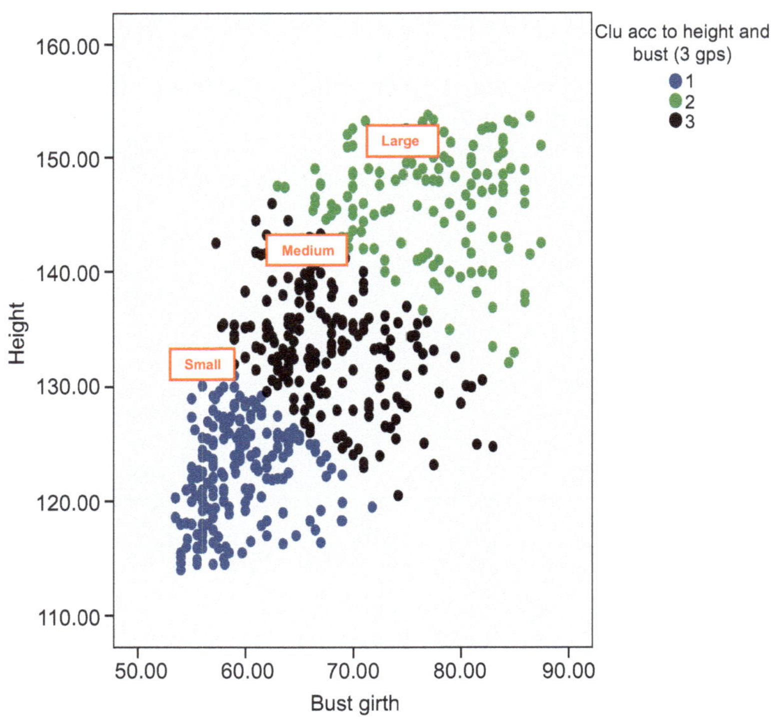

Figure 6.5 Scatter plot of cluster groups according to height and bust for females ages 7–12.

Cluster 2 samples are tall with big bust. Cluster 3 samples are of average height and their bust measurements range from small-to-medium size.

Table 6.6 tabulates the distribution of cluster groups according to age. As shown in the table, the majority are in the small size group (Cluster 1); most in this cluster are 8 years old. The second-highest distribution of the sample ($N=194$) is in the medium size body type (Cluster 3), in which the majority of the sample population are 10 years old. The last group, Cluster 2 ($N=134$), is the large size children and the majority here are 12 years old.

6.4.2.2 Clustering for lower body

For the lower body, the female sample size was divided into three distinct clusters as seen in Table 6.7. A majority of the samples are found in Cluster 2 (medium size) followed by Cluster 1, which is the small group. The last group is Cluster 3, the large group.

Fig. 6.6 shows female samples in three cluster groups divided according to height and hip. In Cluster 1 (small size), samples are mainly found in the short height and narrow hip category. In Cluster 2 (medium size), samples are in the middle of the

Table 6.6 Age versus cluster groups for upper body: females (ages 7–12)

Age	Cluster according to height and bust (three groups)			Total
	Small	*Medium*	*Large*	
7	*58	14	2	74
8	62	33	1	96
9	42	44	12	98
10	17	*58	29	104
11	13	28	41	82
12	5	17	*49	71
Total	197	194	134	525

Note: For this and all similar tables in this section, * indicates the most frequently occurring age(s) in a given cluster. For example, the most frequent age in Cluster 1 is 12 ($N=56$).

Table 6.7 Cluster groups for lower body: females (ages 7–12)

	Cluster 1	Cluster 2	Cluster 3
N	176	196	142
Mean height (cm)	122.6	132.4	145.8
Range (cm)	114.0–132.5	119.5–143.3	133.0–154.0
Mean hip girth (cm)	64.39	73.56	82.24
Range (cm)	58.0–74.0	61.0–91.0	70.0–93.5
Body type	*Small*	*Medium*	*Large*

graph, showing they are in the average height category with small-to-medium size hip. In Cluster 3 (large size), the samples are grouped at the upper right of the graph showing that they belong to the tall height category with wide hip girth.

Table 6.8 shows that a majority of the sample size ($N=196$) fall under Cluster 2, medium size. The majority of the samples in this cluster are 10 years old. The second

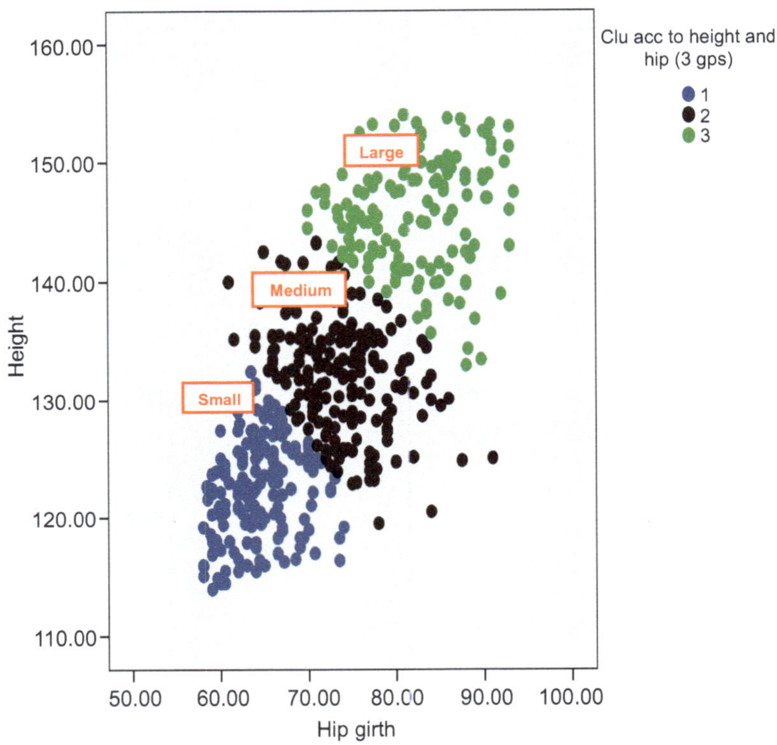

Figure 6.6 Scatter plot of cluster groups according to height and hip for females ages 7–12.

Table 6.8 Age versus cluster groups for lower body: females (ages 7–12)

Age	Cluster according to height and hip (three groups)			Total
	Small	*Medium*	*Large*	
7	53	19	1	73
8	*54	35	2	91
9	33	50	12	95
10	16	*53	35	104
11	14	24	44	82
12	6	15	*48	69
Total	176	196	142	514

Note: For this and all similar tables in this section, * indicates the most frequently occurring age(s) in a given cluster. For example, the most frequent age in Cluster 1 is 12 ($N=56$).

highest distribution of sample size (N=176) is in Cluster 1, small size, wherein the majority are 8 years old. The last cluster group (Cluster 3, N=142) is the large size body type, were the majority are 12 years old.

6.5 The outcome of clustering in a sample population of teenagers

In this section, the tabulation of tables is for the age group of 13–17 years old for different genders. The next sample population group is male (ages 13–17). We begin with the discussion of the upper body clustering.

6.5.1 Male sample size (13–17 years old)

6.5.1.1 Clustering for upper body

Table 6.9 tabulates the distribution of sample groups into three body types. The majority of the sample size fall under Cluster 3, medium body type. The second highest distribution of sample size is in Cluster 1, small size. The fewest samples are in Cluster 2, large size.

Fig. 6.7 gives a visual representation of the sample size showing the samples clustered on the graph of height versus chest girth. Cluster 1 males are mainly short in height with small-to-medium chest measurements. This cluster can be categorized as small size. In Cluster 2, the majority of the subjects are in the upper-right corner of the graph, which means the large size range; they range from short to tall with large chest. It is noted that this cluster has unusually nondifferentiated height range due to the nature of clustering according to two dimensions, length and girth. Cluster 2 seems to be differentiated based on the girth measurements instead of height. Therefore, the distinct characteristic of Cluster 2 is large girth and all children clustered under this group have the largest girth measurements as compared to the other groups. Cluster 3 in contrast has samples that are average-to-tall height with small-to-medium chest measurement, categorized as medium size.

Table 6.10 shows the most frequently occurring ages in each cluster. As mentioned, the majority of the sample population fall under Cluster 3 (medium size) in which

Table 6.9 Cluster groups for upper body: male (ages 13–17)

	Cluster 1	Cluster 3	Cluster 2
N	109	131	80
Mean height (cm)	154.8	167.8	165
Range (cm)	145.9–163.0	159.5–176.7	146.2–176.5
Mean chest girth (cm)	76.7	82.3	95.2
Range (cm)	69.5–89.4	71.6–90.8	88.0–105.5
Body type	*Small*	*Medium*	*Large*

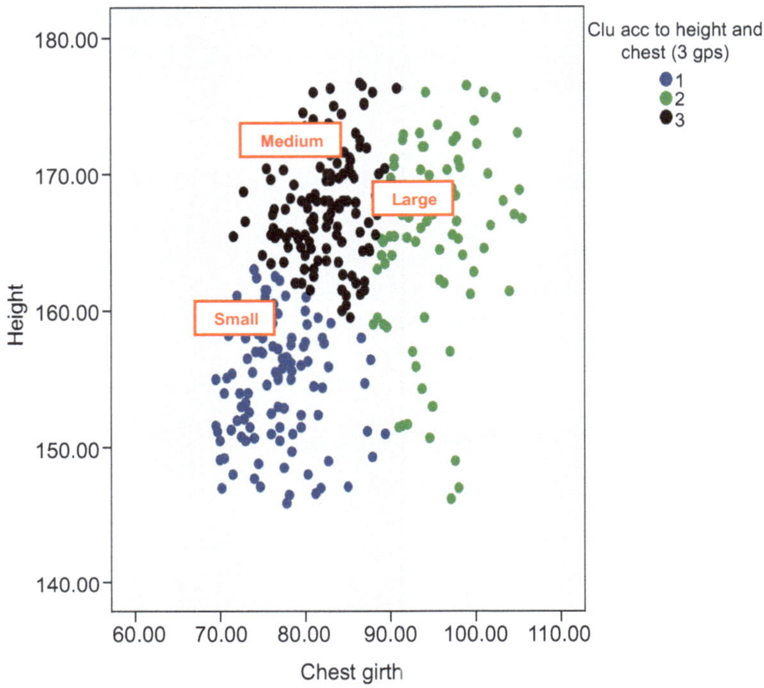

Figure 6.7 Scatter plot of cluster groups according to height and chest for males ages 13–17.

Table 6.10 Age versus cluster groups for upper body: males (ages 13–17)

Age	Cluster according to height and chest (three groups)			Total
	Small	*Medium*	*Large*	
13	*43	12	11	66
14	32	22	18	72
15	19	*44	12	75
16	7	29	*21	57
17	8	24	18	50
Total	109	131	80	320

Note: For this and all similar tables in this section, * indicates the most frequently occurring age(s) in a given cluster. For example, the most frequent age in Cluster 1 is 12 (N=56).

most are 15 years old, whereas in the small size (Cluster 1) the majority are 13 years old. In the large body type (Cluster 2), the distribution of age is fairly even.

6.5.1.2 Clustering for lower body

Using the same k-means cluster technique for lower body dimensions, we successfully extracted three cluster groups (see Table 6.11). The profile of each cluster is defined

Table 6.11 **Cluster groups for lower body: males (ages 13–17)**

	Cluster 3	Cluster 2	Cluster 1
N	111	138	70
Height (cm)	154.6	167.7	165.1
Range (cm)	145.5–163.0	160.0–176.5	146.2–176.5
Chest girth (cm)	79.63	84.7	99.2
Range (cm)	72.6–93.4	73.5–92.8	91.3–107.8
Body type	*Small*	*Medium*	*Large*

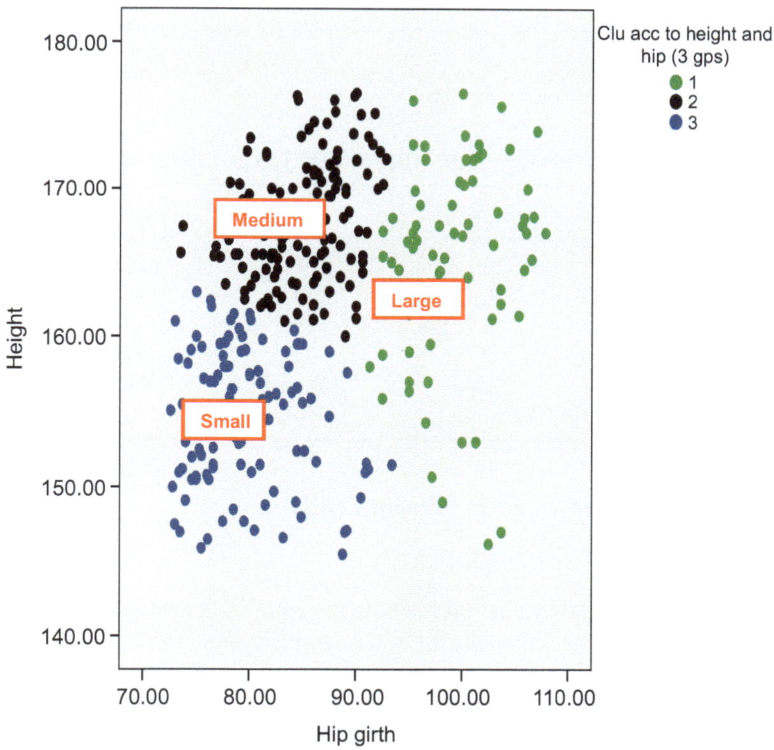

Figure 6.8 Scatter plot of cluster groups according to height and hip for males ages 13–17.

as follows: Cluster 1 (large size) with 70 samples, Cluster 2 (medium size) with 138 samples, and Cluster 3 (small size) with 111 samples. Cluster 2 has the most samples, followed by Cluster 3 and then Cluster 1 with the fewest number of samples.

As seen in Fig. 6.8, Cluster 1 samples are mainly short to tall with wide hip whereas Cluster 2 samples are average to tall with small-to-average hip. Cluster 3 samples are mostly in the short-to-average height, with small-to-average hip size.

The distributions of age in the cluster groups are tabulated in Table 6.12. The majority of the medium body type samples are 15 years old whereas in the small body type the majority are 13 years old. In the large size body type, all the ages are distributed evenly.

Table 6.12 **Age versus cluster groups for lower body: males (ages 13–17)**

Age	Cluster according to height and hip (three groups)			Total
	Small	*Medium*	*Large*	
13	*43	12	12	67
14	33	24	*17	74
15	18	*40	14	72
16	10	33	15	58
17	7	29	12	48
Total	111	138	70	319

Note: For this and all similar tables in this section, * indicates the most frequently occurring age(s) in a given cluster. For example, the most frequent age in Cluster 1 is 12 (N=56).

Table 6.13 **Cluster groups for upper body: females (ages 13–17)**

	Cluster 2	Cluster 3	Cluster 1
N	120	101	80
Mean height (cm)	150.1	159.0	155.2
Range (cm)	142.7–156.0	153.3–164.5	145.0–164.0
Mean bust girth (cm)	76.6	79.6	91.8
Range (cm)	69.5–86.7	70.6–86.7	85.7–98.5
Body type	*Small*	*Medium*	*Large*

6.5.2 Female samples (13–17 years old)

6.5.2.1 Clustering for upper body

Table 6.13 shows the distribution of females (ages 13–17) according to height and bust girth. This table shows that the majority of the sample population fall under Cluster 2, small size. The second highest distribution of sample size is in Cluster 3, medium size. The last cluster group, Cluster 1, is the large size.

The division of female samples (ages 13–17) is depicted in Fig. 6.9. Three clusters are evident. Cluster 1, large size, contains samples that are short to tall with large bust measurements. For Cluster 2, the samples belong to the small size body type with short-to-average height and small-to-average bust. In Cluster 3, medium size, the samples are average to tall with small-to-average size bust.

As can be seen from Table 6.14, the small size cluster (Cluster 2) has most of the sample size, and the majority of this cluster are age 13. Ages are evenly distributed in the medium body type, whereas females age 16 are mostly found in the large size cluster group.

6.5.2.2 Clustering for lower body

Table 6.15 shows the profile of each cluster for this sample size according to lower body key dimensions. The majority of the sample size falls under Cluster 2, the small

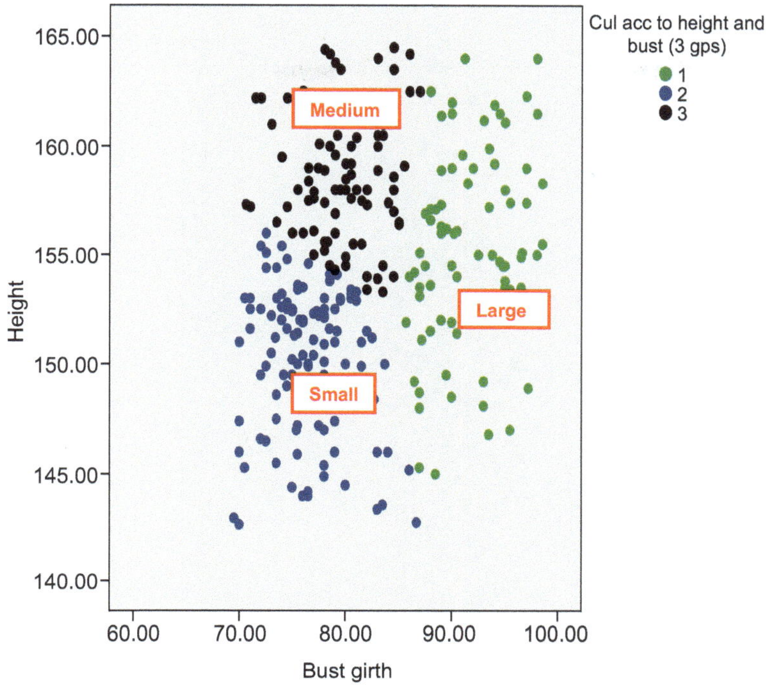

Figure 6.9 Scatter plot of age groups according to height and bust for females ages 13–17.

Table 6.14 Age versus cluster groups for upper body: females (ages 13–17)

Age	Cluster according to height and bust (three groups)			Total
	Small	*Medium*	*Large*	
13	*36	14	15	65
14	22	24	18	64
15	25	19	13	57
16	17	20	*23	60
17	20	*24	11	55
Total	120	101	80	301

Table 6.15 Cluster groups for lower body: females (ages 13–17)

	Cluster 2	Cluster 3	Cluster 1
N	126	102	64
Mean height (cm)	150.4	158.9	155.8
Range (cm)	143.4–157.3	152.9–164.5	142.8–164.0
Mean hip girth (cm)	84.1	88.3	99.3
Range (cm)	76.0–92.5	81.5–94.2	92.5–106.5
Body type	*Small*	*Medium*	*Large*

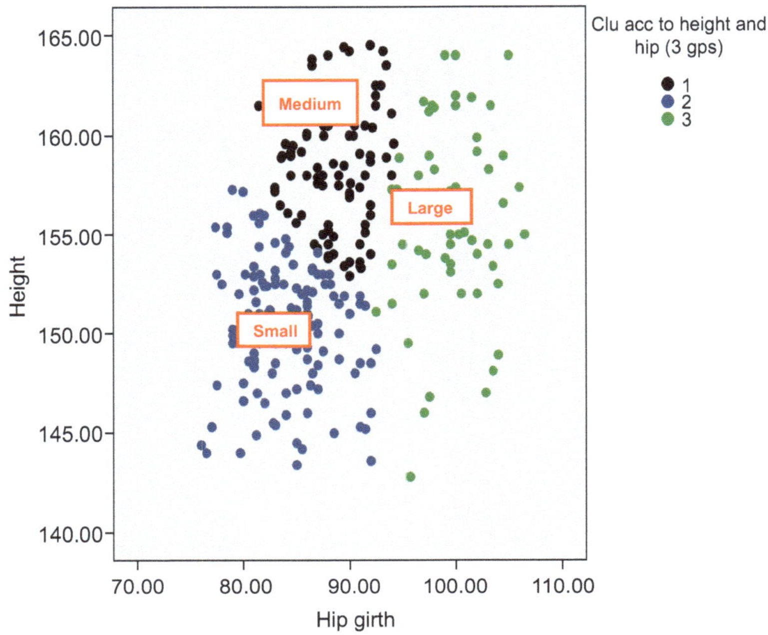

Figure 6.10 Scatter plot of age groups according to height and hip for females ages 13–17.

Table 6.16 Age versus cluster groups for lower body: females (ages 13–17)

Age	Cluster according to height and hip (three groups)			Total
	Small	*Medium*	*Large*	
13	*35	16	10	61
14	23	*23	16	62
15	26	*23	8	57
16	18	20	*21	59
17	24	20	9	53
Total	126	102	64	292

size. The second highest distribution of sample size is in Cluster 1, the medium size. The last cluster size, Cluster 3, is the large size.

Fig. 6.10 illustrates the distribution of sample size according to height and hip girth separated into three different groups. Small size samples differ from medium size samples in height range. Both cluster groups have the same small-to-average hip measurements, but the small size has small-to-average height whereas the medium size has medium-to-tall height. Large size samples are short to tall with large hip girth.

Table 6.16 depicts the distribution of samples according to age and body type. The majority of small size samples are age 13 marked with *. For the medium size cluster,

Table 6.17 **Anthropometric data characteristics**

Body type	Body characteristics
Small	Short height with small girths (see Fig. 5.15)
	Short height with small-to-large girths (see Fig. 5.16)
Medium	Average height with average girths (see Fig. 5.17)
	Average-to-tall height with small girths (see Figs. 5.15 and 5.16)
Large	Short-to-tall height with large girths (see Fig. 5.15)
	Average-to-tall height with large girths (see Fig. 5.16)
	Tall height with small-to-large girths (see Fig. 5.17)

all ages are fairly evenly distributed, although ages 14 and 15 seemed to be with the highest number marked with *. The last cluster group (Cluster 3) is the large size, wherein the dominant age is 16 marked with *.

6.6 Body characteristics in male and female groups

This study has demonstrated a very well-defined cluster groups using the partitioning k-means method. Moreover, this study also confirms distinct results when both height and girth variables are used as the key dimensions (primary control dimensions), as compared to common practice where height is the primary control dimension followed by girth dimensions as the secondary control dimensions [5,7–10,14].

The most striking result that emerges from this section is the discovery of the different body characteristics in the samples when they are grouped according to key dimensions. These body characteristics are depicted in Table 6.17 and Figs. 6.11–6.13. The results suggest that there are seven characteristics observed in the three different body types (small, medium, and large size).

Figs. 6.11–6.13 illustrate the seven body characteristics mentioned in Table 6.17. There are two types of small sizes, two types of medium sizes, and three types of large sizes. For example, in Fig. 6.15 the small size is represented by small girth and short height. The large size is depicted by large girth and short-to-tall height, and finally the medium size is signifying the average-to-tall height with small girths. On the other hand, Figs. 6.11 and 6.12 depict the other body characteristics described in Table 6.17.

The total yield for the whole anthropometric data set is 24 cluster groups (12 for upper body and 12 for lower body) for ages 7–17, which contain any one of the body characteristics described in Table 6.17. This result is different from previous studies in which 36 cluster groups were obtained for a similar age range (6–18 years old) [5]. Figs. 6.11–6.13 below illustrate the main characteristics of the different age groups, 7–12 and 13–17. In another recent study, the samples are grouped into five clusters.

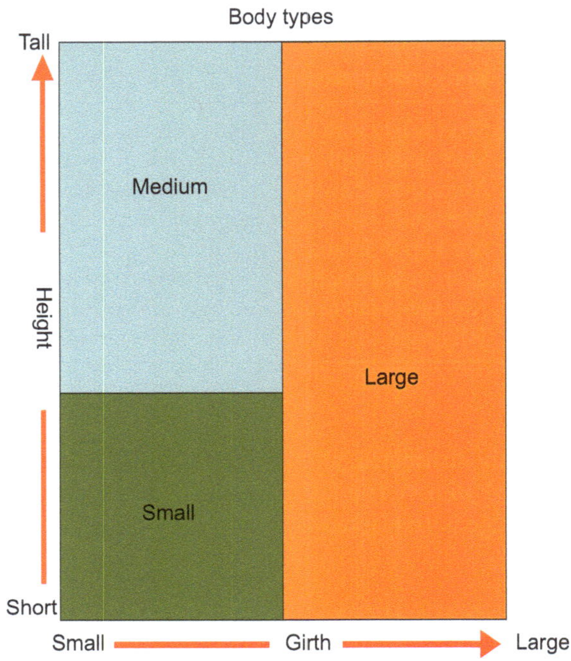

Figure 6.11 Body characteristics type 1.

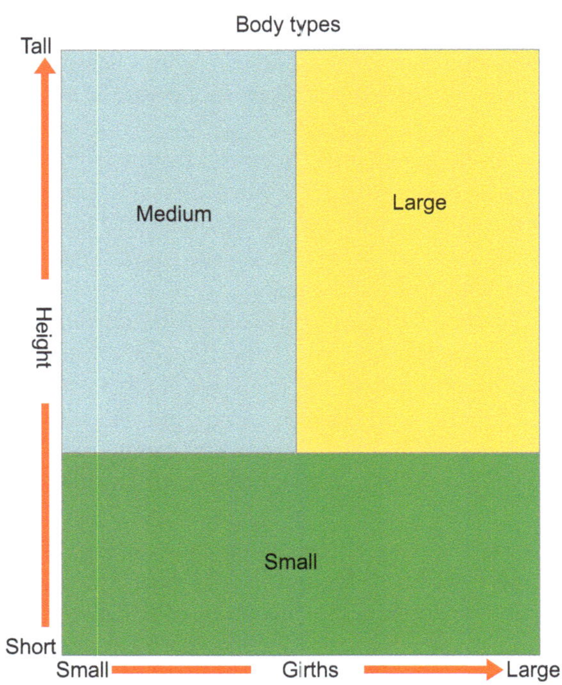

Figure 6.12 Body characteristics type 2.

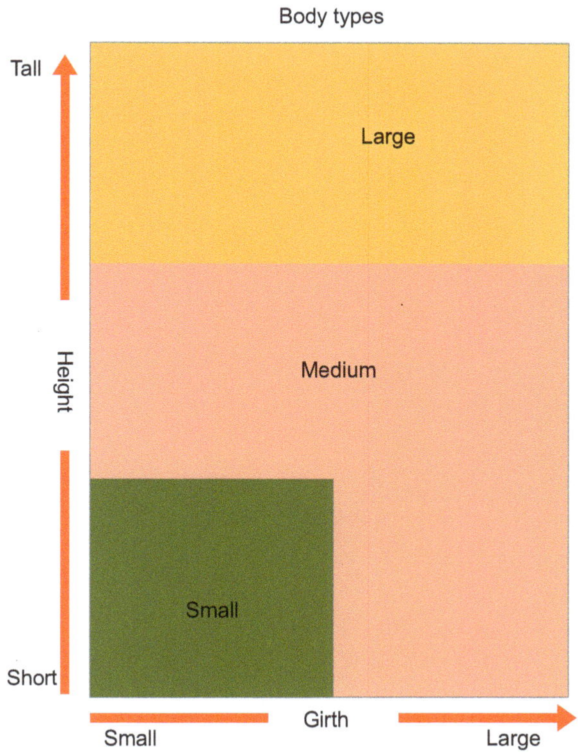

Figure 6.13 Body characteristics type 3.

6.7 Summary

In Sections 6.5.1 and 6.5.2, the results of cluster analysis are reviewed for males and females ages 7–12. In this section, Figs. 6.14 and 6.15 summarize the final body characteristics extracted via k-means cluster technique for both genders (male and female) and body regions (upper and lower body).

6.7.1 Cluster sampling technique for sample group ages 7–12

On the right of Fig. 6.14 are the body characteristics for females, and on the left of the figure are the body characteristics of males. As can be seen, this study confirms that there are two different cluster trends that describe six different body characteristics in females and males of the same age group (7–12). The same cluster trend can be seen for both upper and lower body for each gender. This is in agreement with the earlier observations in Chapter 4, which proves that male and female anthropometric data differ significantly. Moreover, the same observation is also seen in other previous studies where differences of anthropometric data are found in both genders. However, different cluster trends found in both genders of the same age range have not previously been described. For example, in other research it was found that there is a linear growth in all the cluster groups [5].

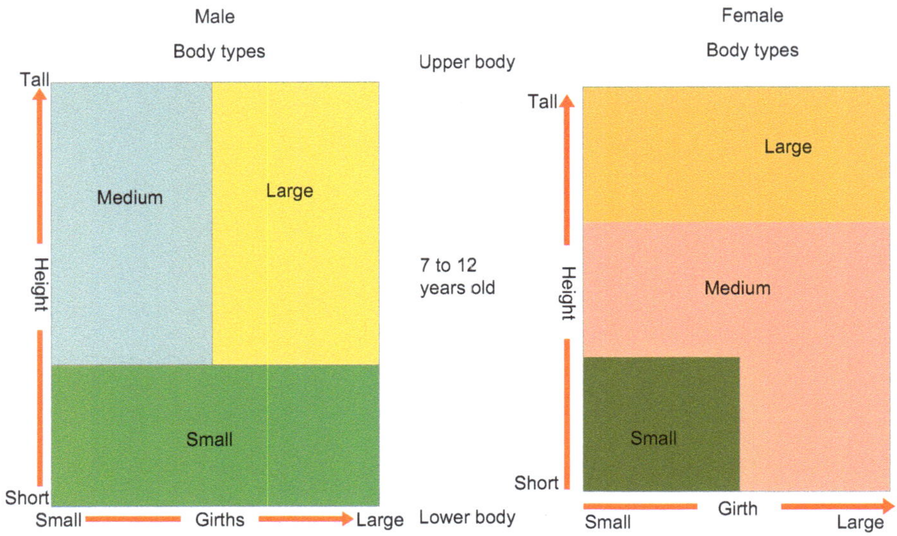

Figure 6.14 Body characteristics of males and females ages 7–12.

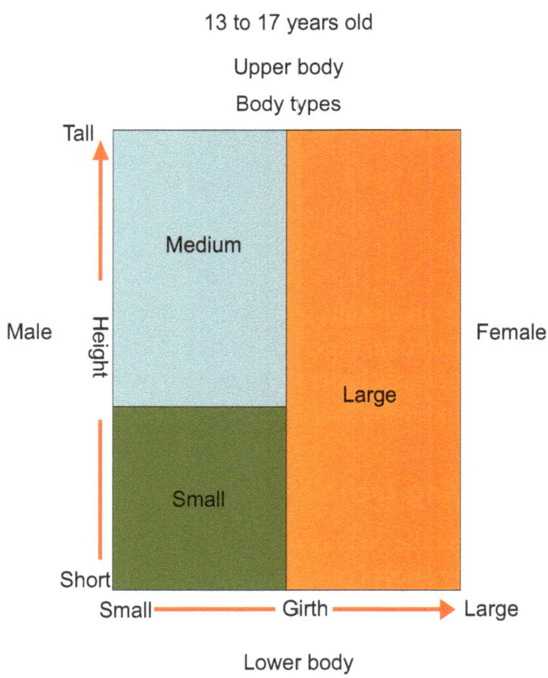

Figure 6.15 Body characteristics of males and females ages 13–17.

In typical sizing research of three clusters, linear trends are observed that are positively correlated, that is, if the height is tall, the girth is also big and vice versa, indicating that the body types are small, medium, and large [5,7,8]. The same observation is shown in female samples (Fig. 6.5). In contrast, the trend of body types is not linear for the male sample population. This implies that if someone is tall, it does not necessarily follow that the girths are large and vice versa. This cluster trend implies that the understanding of different characteristics in body relationship determines the effectiveness of sizing system development. Furthermore, in this age group (7–12), the obvious pattern is that the younger age group belongs to the smaller size body type. For example, the majority of children ages 7 and 8 are often in small size, ages 9 and 10 in medium size, and ages 11 and 12 always in the large size group. For younger children (ages 7–12) their height is differentiated in each of the groups as short, average height, and tall. Therefore, by knowing the characteristics of body dimensions, an apparel manufacturer is then able to manufacture clothing that fits the human body variations.

6.7.2 Cluster sampling technique for sample group ages 13–17

In this Section 6.7.2, Fig. 6.14 indicates that in this age group, only one cluster trend is observed with three different body characteristics. A distinct finding from this age group is the body characteristic dictated by the girth measurement, which is not found in the younger age group (7–12 years old). As can be seen from Fig. 6.15, the large group size belongs to teenagers with large girth measurements. Height is not differentiated in the large size, which means one can be short or tall but if they have large girth they go under the same size category. This size is mainly for children with big girths.

This finding shows that the size found from the analysis can be categorized according to body shape instead of height category. For ages 13–17, the children are divided to three different body characteristics, namely petite (short and small girth), slim (medium height to tall with medium girth), and large size (short to tall with big girth).

It can be concluded that body shapes are more obvious in this age group (13–17), where petite, slim, and large sizes are found. In the age group 7–12, it is obvious the group is being divided according to height characteristics.

Although the same body characteristics are found in both genders, this does not mean that both genders are the same. The profiles of each characteristic differ significantly. This is the purpose of conducting the decision tree technique on the cluster groups—to check the percentage of correctness for each sample and to extract the profile for each sample. The same cluster trend observation can be seen in previous literature [6].

In terms of age, there are also distinct differences. This study confirms that each age is found in each cluster group. However, in every cluster group a dominant age trend can be seen. For example, in terms of both upper and lower body, females ages 7 and 8 are clustered in the small size body type, ages 9 and 10 in the medium size, and ages 11 and 12 in the large size. For the male population, in the upper body, ages 7 and 8 are clustered in small type, ages 9 and 10 mostly in the medium size, and ages 11 and 12 in the large body type. In the lower body, ages 7 and 8 are in the small size, ages 9, 10, and 11 are mostly in the medium size, and most ages 11 and 12 are in the large size. Males who are 11 years old fall equally into two different figure types: medium and large. Lastly, the analysis in this section proves that wide variations of figure types do exist for different age groups (7–12 and 13–17) and genders (male and female). This indicates that a separate sizing system is needed for each of these groups.

The next chapter reviews the decision tree analysis that is used to validate the cluster results and to classify the body types before the final sizing system can be developed.

References

[1] O. Abdali, P. Eric, R. Marc, Exploring anthropometric data through cluster analysis, Int. J. Consumer Stud. 32 (2004) 276–284.
[2] Z. Peter, Cluster analysis in sociometric research: a pattern-oriented approach to identifying temporally stable peer status groups of girls, J. Early Adolesc. 27 (2007) 90–114.
[3] H.L. Viktor, E. Paquet, H. Guo, Measuring to fit: virtual tailoring through cluster analysis and classification, in: Presented at 17th International European Conference on Machine Learning and the 10th European Conference on Principles and Practice of Knowledge Discovery in Database, September 18–22, Berlin,Germany, 2006.
[4] T. Raykov, G.A. Marcoulides, An Introduction to Applied Multivariate Analysis, 2008.
[5] M.J. Chung, H.F. Lin, J.J. Mao, J. Wang, The development of sizing systems for Taiwanese elementary-and high-school students, Int. J. Ind. Ergon. 37 (2007) 707–716.
[6] C. Hsu, Data mining to improve industrial standards and enhance production and marketing: an empirical study in apparel industry, Exp. Syst. Appl. 36 (2009) 4185–4191.
[7] H.C. Hsu, H.F. Lin, M.J. Wang, Developing female size charts for facilitating garment production by using data mining, J. Chin. Inst. Ind. Eng. 24 (2007) 245–251.
[8] A.S. Ariadurai, T.P.G. Nilusha, M.R. Dissanayake, An anthropometric study on Sri Lankan school children for developing clothing sizes, J. Soc. Sci. 19 (2009) 51–56.
[9] C.H. Hsu, M.J.J. Wang, Using innovative technology to establish sizing systems, Int. J. Innov. Learn. 2 (2005) 233–245.
[10] D. Gupta, B.R. Gangadhar, A statistical model for developing body size charts for garments, Int. J. Cloth. Sci. Technol. 16 (2004) 458–469.
[11] A. Petrova, Creating sizing systems, in: S.P. Ashdown (Ed.), Sizing in Clothing Developing Effective Sizing Systems for Ready-to-Wear Clothing, Woodhead Publishing, Cambridge, 2007, pp. 57–82.
[12] J.C. Yoon, C.R. Jasper, Garment-sizing systems: an international comparison, Int. J. Cloth. Sci. Technol. 5 (1993) 28–37.
[13] D. Gupta, N. Garg, K. Arora, N. Priyadarshini, Developing body measurement charts for garment manufacture based on a linear programming approach, J. Text. Appar. Technol. Manage. 5 (2006) 1–13.
[14] H.F. Lin, C.H. Hsu, M.J.J. Wang, An application of data mining technique in Engineering to facilitate production management of garments, in: Presented at Proceedings of the 11th WSEAS International Conference on Computers, Greece, 2007.

Classifying the body shapes of children and teenagers for improved apparel fit

7

7.1 Introduction: classification of children and teenagers

This chapter presents Step 4, the last step from sizing development, which is the classification of the cluster groups to gain profiles for each group for sizing system development as shown in Fig. 7.1. As can be seen, there are two main steps used to verify the cluster groups and also to profile the cluster groups. The verification of the groups will validate each cluster group before it is profiled into the correct cluster group. The classification profile will use both the dependent variables and independent variables to obtain the profile of the body classification.

The objective of this analysis is to achieve the classification profile of the cluster groups. The methodology of this chapter is to use the classification and regression tree technique (CRT). The goal of this chapter is to show how the cluster groups are being classified with a very clear guideline as to where the groups should belong.

In this chapter, Section 7.2 presents the classification methodology. Section 7.3 will provide the results of classification for variations in body shapes for children ages 7–12 and teenagers ages 13–17. Section 7.4 will show how the classification decision tree technique profiles the body shapes for a better sizing system. Section 7.5 will give the conclusions of this chapter.

7.2 Classification methodology

The decision tree analysis is a data mining technique. It is an effective technique for classification [1]. For this study, the CRT technique was used to verify and classify the sample population according to cluster groups; CRT was chosen because the data are continuous, and CRT is appropriate for continuous data. The profile of the tree is useful when interpretation of the data set is required. By doing the classification analysis, important variables can be obtained and a simple profile can easily be extracted from the tree diagram [2].

The extracted cluster profile is used as a guide for dividing the cluster groups into different body types for the development of a sizing system. This classified profile found from the tree diagram clearly describes the profile of each cluster group. Moreover, each profile describes the rules of how to split the data into the correct cluster groups [2]. Cases can be predicted and the percentage of correctness can be

Clothing for Children and Teenagers. http://dx.doi.org/10.1016/B978-0-08-100226-1.00007-2
Copyright © 2016 Elsevier Ltd. All rights reserved.

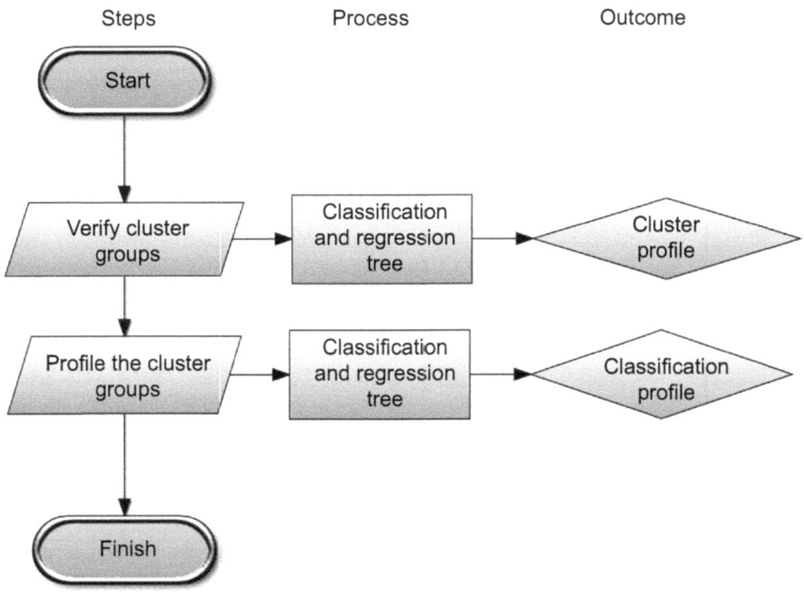

Figure 7.1 Decision tree data mining process.

established for every sample population. The CRT method attempts to maximize within-node homogeneity [1]. The CRT technique first puts the preprocessed data into the root of a tree; it then performs a series of classifications to find the best separating points, thus forming a tree-shaped structure [3].

The average correctness should be in the range of 80% to show that the cluster groups are significant and efficient. This ensures selection of key dimensions that yield the highest coverage of sizing systems. The decision tree is also used to profile the cluster groups. The dependent variable was the cluster groups and the independent variables used are the key dimensions. The key dimensions are used as the independent variables to verify the correctness of using the key dimensions to cluster the groups. Thus, the decision tree technique is employed for two reasons: to verify the cluster groups and to determine the profile of each cluster group for the development of sizing system.

The CRT predictive modeling technique employed for this study depicts the rules for separating the data into significant groups based on the three cluster groups established from cluster analyses in the previous section.

Moreover, these results verify that the samples are classified into their own cluster groups with high accuracy. It also suggests that the accuracy of classification is excellent in all four sample groups, with more than 80% accuracy for ages 7–12 and above 90% for ages 13–17. Hence, this confirms that the sample groups are classified correctly into cluster groups using k-means technique. The profiles of each split are given in the following paragraphs.

For this section and all of the following sections, the same dependent and independent variables are used to classify the cluster groups. The dependent variables are the cluster groups for each individual sample group. The independent variables are the primary control variables (key dimensions) of each sample group specific to either upper body or lower body. The same observations are seen in all sample groups of both upper and lower body where the rule of this tree model is as follows: the maximum depth of the tree extends two levels beneath the root node except in females ages 7–12 of lower body where it is observed that the maximum depth extends to three levels beneath the root node. The minimum sample at the parent node is 100 and the child node is 50, which is the same for all of the sample groups. The same observation is also seen in the predictive variables where height is selected as the most important variable for all sample groups ages 7–12. In comparison, for ages 13–17, the girth variables are identified as the predictive variables.

7.3 Classification results for variations in body shapes

As seen in Chapter 6, the cluster analysis technique has shown how each sample group is segmented into three different groups as shown in the Chapter 6 tables (eg, Table 6.1 reproduced here) in which the range was given for both the key dimensions. For example, the key dimensions for upper body are the height and chest girth. Here the group is divided into three main groups based on the key dimensions, and the range of the key dimensions is given such as height 114.5–132.7 cm and 55.0–80.4 cm for chest.

It can be seen from the cluster groups that the size ranges overlap one another, which makes it difficult to determine the sizes. The cluster analysis method gives the profile of only the body shapes based on the profiling of the key dimensions, which means that the sample population is divided according to the key dimensions and each group has the same body sizes. If a person is 127 cm and has a chest girth of 73 cm, he can be in size small, medium, and large. This makes choosing a size confusing. Therefore, the next step to determine the sizes unmistakably is to do the data mining using the decision tree technique in which each cluster is profiled according to their own body sizes.

7.3.1 Decision tree for males (ages 7–12), upper body

Table 7.1 illustrates the results of data mining according to cluster groups obtained from previous analyses. Columns labeled 1, 2, and 3 depict the cluster groups that show the number of samples in respective cluster groups. The last row of each cluster group indicates the percentage of that cluster group in comparison to other cluster groups. It is revealed that 37.5% of the samples are grouped in Cluster 1, with the majority in Cluster 1 with 193 samples, Cluster 2 with 12 samples, and Cluster 3 with 6 samples. In addition, 41.6% represent Cluster 2 and another 21.0% represent Cluster 3. This means that majority of these samples are found in Cluster 2.

These details are seen in every regression table in the subsequent sample groups. However, the description is not given but instead only the overall percent of correctness for each sample group is emphasized as the goal of decision tree is

Table 7.1 Regression tree validation, upper body (males ages 7–12)

Observed	Predicted			
Male (7–12)	**1**	**2**	**3**	**Percent correct (%)**
1	193	4	0	98.1
2	12	222	0	94.9
3	6	8	118	89.4
Overall percentage (%)	37.5	41.6	21.0	94.1

to verify the efficiency of cluster groups. Furthermore, the details of each cluster group are self-explanatory.

Table 7.1 also illustrates the percentage of correctness, which demonstrates the efficiency when classified according to their respective cluster groups. Overall, this result indicates that samples are correctly classified in their own groups, which shows an excellent average of 94.1%. The specific percentage of correctness for each cluster group is given as these: Cluster 1 has 98.1%, Cluster 2 has 94.9%, and Cluster 3 has 89.4% correctness.

Fig. 7.2 shows that height is also the only significant predictor for the small size group (Cluster 2). Samples with height less than 129.6 cm are most likely to be grouped under the small size. For Cluster 1 and Cluster 3, the next best predictor is chest girth. These two cluster groups have the same minimum height range but different girth ranges. All samples in Cluster 3 are definitely large size with height greater than 129.6 cm and chest girth greater than 74.6 cm. On the other hand, those with the same height (more than 129.6 cm) but with less than 74.6 cm chest girth are most likely to be grouped under the medium size. In addition, the profiled rules of each sample group are shown in Table 7.2.

7.3.2 Decision tree for males (ages 7–12), lower body

Table 7.3 shows the verification of cluster groups with correctness of 78.0% for Cluster 1, 95.2% for Cluster 2, and 82.9% for Cluster 3. The results for this sample show an average correctness of 85.4%, which is considered good.

As shown in Fig. 7.3, height is the only significant attribute for small size samples. The samples that are shorter than 128.6 cm are most likely to be grouped under small body size. In addition, it can be seen that height and hip girth are the significant attributes for medium and large size samples. Samples that are more than 128.6 cm tall are most likely to be grouped under large size with hip girth more than 76.9 cm. Lastly, the samples are likely to be in the medium size group if they are more than 128.6 cm in height with less than 76.9 cm hip measurements. The profile for each body size is illustrated in Table 7.4.

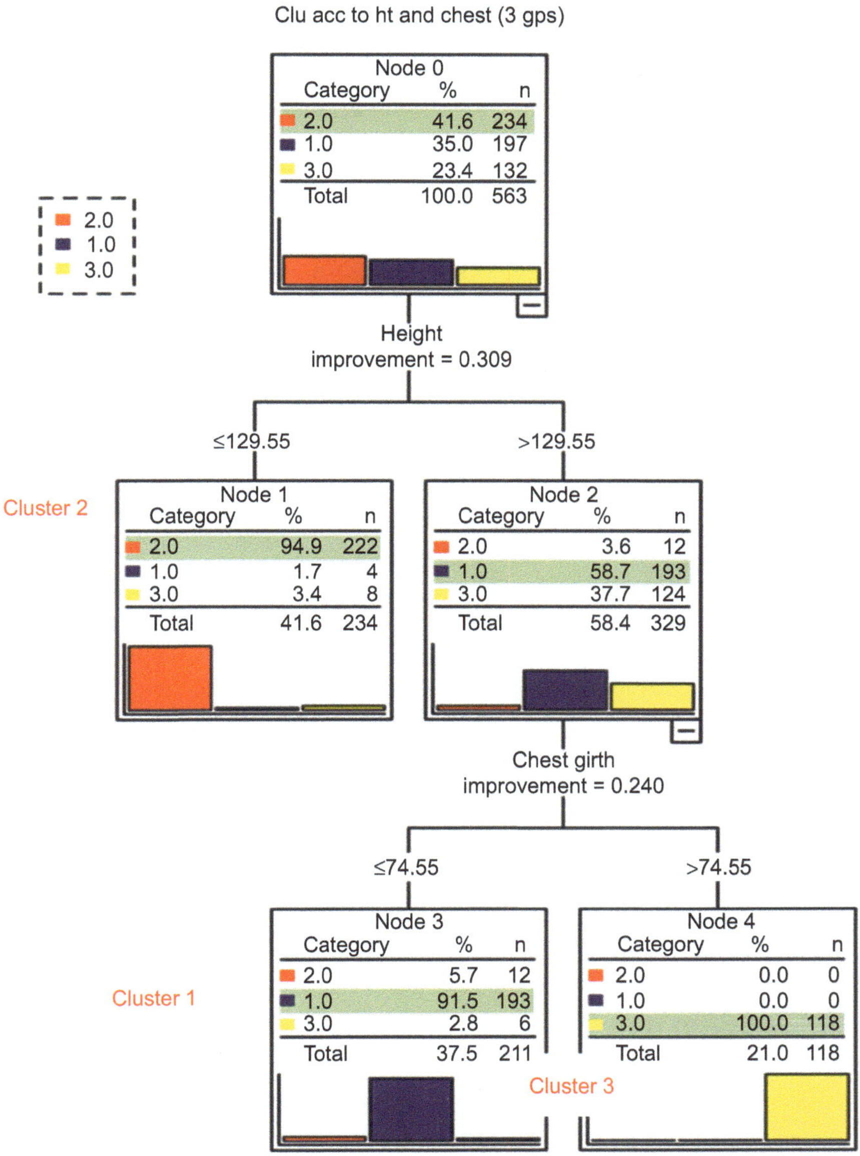

Figure 7.2 Decision tree for males (ages 7–12), upper body.

Table 7.2 Profile of upper body types for males ages 7–12

Node	Body type	Classified rule	*N*
1	Small	Height ≤ 129.6 cm	222
2 and 3	Medium	Height > 129.6 cm and chest girth ≤ 74.6 cm	193
2 and 4	Large	Height > 129.6 cm and chest girth > 74.6 cm	118

Table 7.3 **Regression tree validation, lower body (males ages 7–12)**

Observed	Predicted			
Males (7–12)	1	2	3	Percent correct (%)
1	128	1	35	78.0
2	0	178	9	95.2
3	14	19	160	82.9
Overall percentage (%)	26.1	36.4	37.5	85.4

7.3.3 Decision tree for females (ages 7–12), upper body

For females ages 7–12, the results of CRT analysis are as follows. As shown in Table 7.5, the decision tree classified the cluster groups with correctness of Cluster 1, Cluster 2, and Cluster 3 as 88.8%, 90.3%, and 86.6%, respectively. The final overall percentage for this cluster group is 88.4%.

As seen from Fig. 7.4, the samples are first split into two height ranges. One group's height is more than 129.4 cm and the other's is less than 129.4 cm. Those who are more than 129.4 cm tall with bust measurement greater than 62.3 cm are most likely to be grouped under the small size (Cluster 1). Those with the same height but with bust less than 62.3 cm are likely to be grouped under the medium size (Cluster 3). Those samples with height greater than 129.4 cm and less than 139.5 cm are more likely to be grouped under the medium size (Cluster 3). Lastly, those who are taller than 139.7 cm are more likely to be grouped under the large size (Cluster 2). The tree validates the significance of the control variables. The profiles of the clusters shown in Table 7.6 are as follows:

7.3.4 Decision tree for females (ages 7–12), lower body

Decision tree analysis classified these cluster groups with a correctness of 87.4% as a whole. The correctness of classification for Cluster 1 is 99.4%, for Cluster 2 is about 71.9%, and for Cluster 3 is about 90.9%, as shown in Table 7.7.

As shown in Fig. 7.5, samples are first divided into two height measurements, greater than and less than 139.7 cm. Those who are more than 139.7 cm tall are most likely to be grouped under the large size, and height seems to be the only significant characteristic for large size people. Those with height less than 139.7 cm and hip less than 67.4 cm are likely to be grouped in the small size. Those with height greater than 126.5 cm and less than 139.7 cm with hips greater than 67.4 cm are likely to be clustered under the medium size. Those with height less than 126.5 cm and hips greater than 67.5 cm are also likely to be grouped under medium size. Table 7.8 summarizes the profiles.

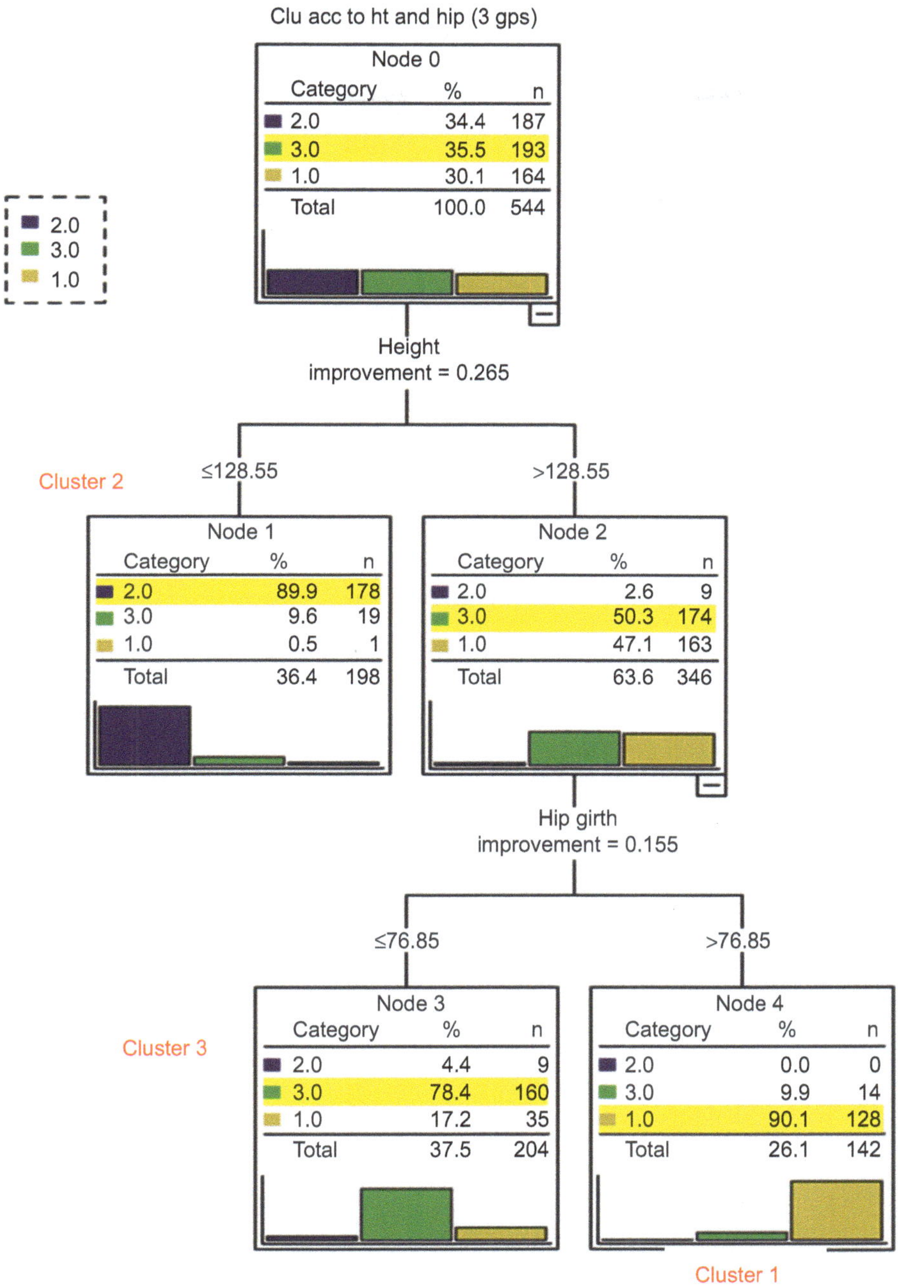

Figure 7.3 Decision tree for males (ages 7–12), lower body.

Table 7.4 **Profile of lower body types for males ages 7–12**

Node	Body type	Classified rule	*N*
1	Small	Height ≤ 128.6 cm	178
2 and 3	Medium	Height > 128.6 cm and chest girth ≤ 76.9 cm	160
2 and 4	Large	Height > 128.6 cm and chest girth > 76.9 cm	128

Table 7.5 Regression tree validation, upper body (females ages 7–12)

Observed	Predicted			
Female (7–12)	1	2	3	Percent correct (%)
1	175	0	22	88.8
2	0	121	13	90.3
3	3	23	168	86.6
Overall correctness (%)	33.9	27.4	38.7	88.4

7.3.5 Decision tree for males (ages 13–17), upper body

As can be seen from Table 7.9, the decision tree for this set of measurements verified the cluster groups with correctness of 93.6% for Cluster 1, 96.3% for Cluster 2, and 95.4% for Cluster 3. The overall correctness for this cluster group is excellent at 95.1%.

The result of the tree (Fig. 7.6) shows that chest is the most important variable in this group. The samples were divided into two chest measurement ranges. Those with chest measurements greater than 88.9 cm are most likely to be classified in the large size. Those with chest girth measurements less than 88.9 cm and more than 161.1 cm tall are most likely to be in the medium size group, whereas those who have the same chest measurement but are less than 161.6 cm in height are mostly likely to be in the small size category. The summary of this sample group is depicted in Table 7.10.

7.3.6 Decision tree for males (ages 13–17), lower body

Table 7.11 depicts the result of the CRT technique employed for this set of measurements. It verified the cluster groups with correctness of 75.7% for Cluster 1, 97.8% for Cluster 2, and 95.5% for Cluster 3, as shown in Table 7.11. The overall correctness for this cluster group is excellent with 89.7%.

The results of the tree diagram (Fig. 7.7) show that height is the most important variable in this model. This finding is not the same in other studies as height does not show any differences. Samples that are less than 161.2 cm are most likely to be grouped under small size (Cluster 3). In contrast, samples with height more than 161.2 cm can be found in large (Cluster 1) and medium size (Cluster 2). Those who are more than 161.2 cm tall with hip girth of less than 93.1 cm are most likely to be in the medium size group. In comparison, those who are tall and have more than 93.1 cm hip girth definitely belong to the large size. The profile for this sample group is shown in Table 7.12.

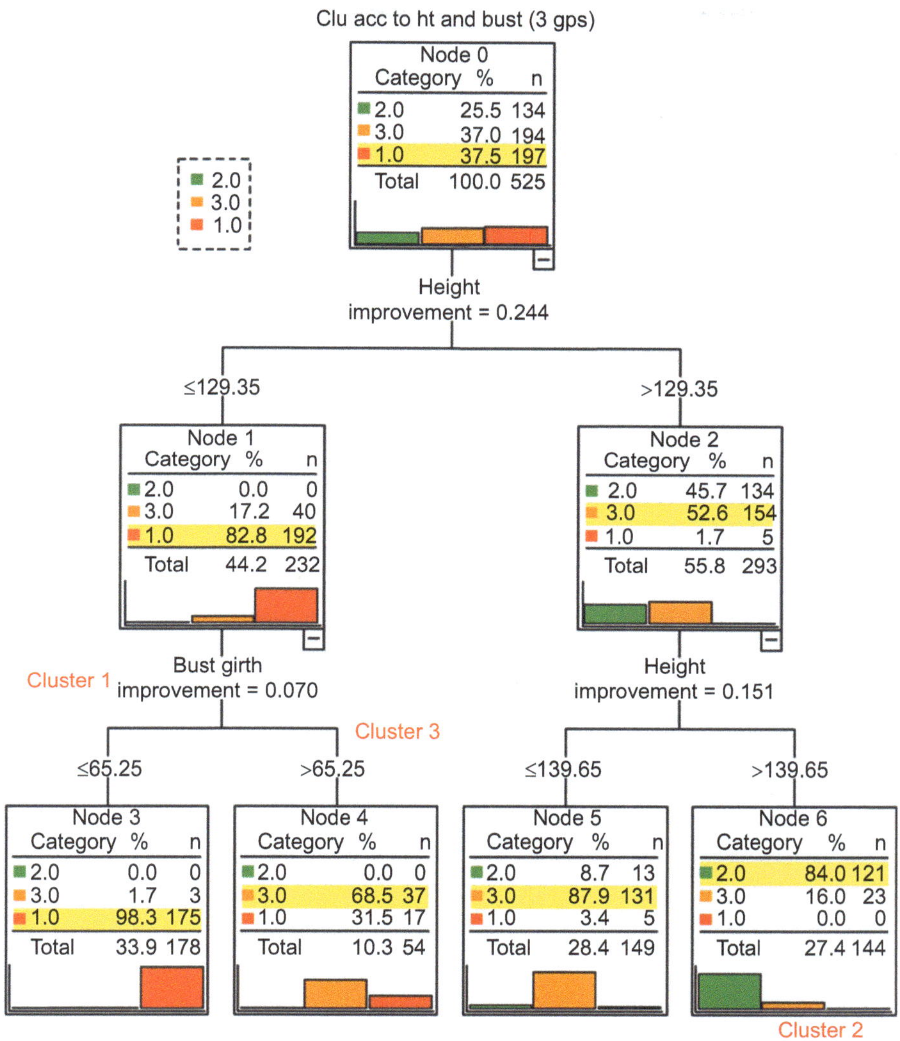

Figure 7.4 Decision tree for females (ages 7–12), upper body.

Table 7.6 **Profile of upper body types for females ages 7–12**

Node	Body type	Classified rule	*N*
1 and 3	Small	Height ≤ 129.4 cm and bust girth ≤ 65.3 cm	175
2 and 5	Medium	Height > 129.4 cm and < 139.7 cm	131
1 and 4	Medium	Height < 129.4 cm and bust girth > 65.3 cm	37
2 and 6	Large	Height > 139.7 cm	121

Table 7.7 **Regression tree validation, lower body (females ages 7–12)**

Observed	Predicted			
Female (7–12)	1	2	3	Percent correct (%)
1	175	1	0	99.4
2	42	141	13	71.9
3	0	13	129	90.9
Overall percentage (%)	42.2	30.2	27.6	87.4

7.3.7 Decision tree for females (ages 13–17), upper body

As shown in Table 7.13, the decision tree verified the cluster groups for this population with correctness of 97.5% for Cluster 1, 90.8% for Cluster 2, and 97.0% for Cluster 3; the overall correctness for this cluster group is excellent at 94.7%.

The results of the tree (Fig. 7.8) show that bust girth is the only predictor for the large size. The female samples in this group were split into two bust measurements. Those with bust measurements greater than 86.3 cm are most likely to be clustered under the large size. Those with bust measurement less than 86.3 cm and more than 161.4 cm tall are likely to be grouped as the medium size. Those less than 154 cm tall with bust less than 86.2 cm are likely to be grouped under small size. The profile of the clusters is shown in Table 7.14 as follows:

7.3.8 Decision tree for females (ages 13–17), lower body

The decision tree for this set of measurements verified the cluster groups with correctness of 96.1% for Cluster 1, 85.7% for Cluster 2, and 98.4% for Cluster 3 (see Table 7.15). The overall percentage for this cluster group is excellent at 92.1%.

Fig. 7.9 shows that hip girth is the most important control variable in this group. The other significant variable is height. The samples that are more than 93.8 cm tall are mostly grouped as large size. Those taller than 153.3 cm and with hip girth less than 86.3 cm are most likely to be grouped as medium size. Those who have hip girth less than 93.75 cm and are less than 153.3 cm tall are most likely to be in the small size group. The profiles of the cluster groups are depicted in Table 7.16.

7.4 The classification rules based on gender and age groups

Tables 7.17 and 7.18 are the profiles of the sample population according to different age group and genders for upper and lower body. These tables depict the characteristics of each body type that denote the sizes (small, medium, and large). Characteristics

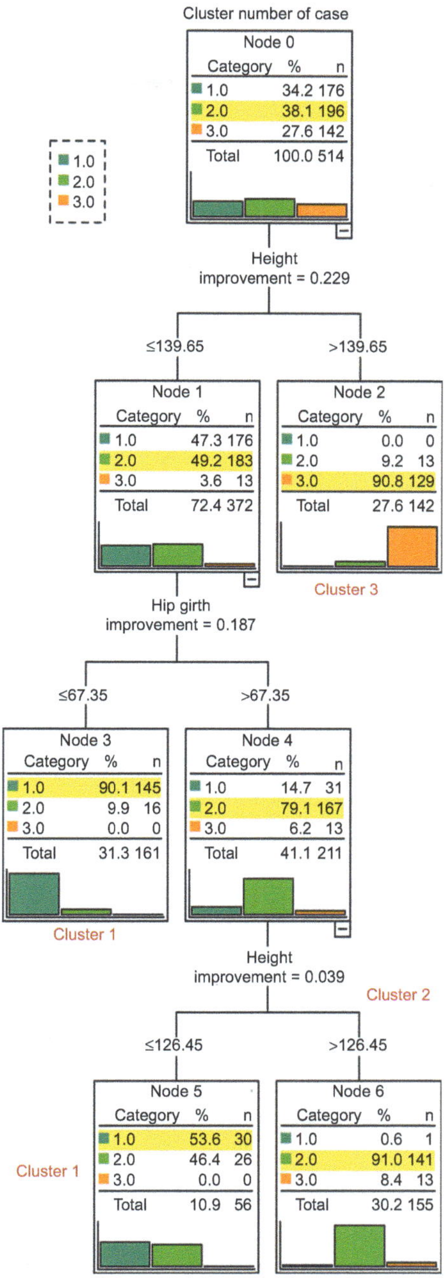

Figure 7.5 Decision tree for females (7–12 years old), lower body.

Table 7.8 Profile of lower body types for females ages 7–12

Node	Body type	Classified rule	*N*
1 and 3	Small	Height ≤ 139.7 cm and hip ≤ 67.4 cm	145
1, 4 and 5	Small	Height < 126.5 cm and hip > 67.4 cm	30
1, 4 and 6	Medium	Height > 126.4 cm and ≤ 139.7 cm and hip > 67.4 cm	141
2	Large	Height > 139.7 cm	129

Table 7.9 **Regression tree validation, upper body (males ages 13–17)**

Observed	Predicted			
Male (13–17)	**1**	**2**	**3**	**Percent correct (%)**
1	102	1	6	93.6
2	2	77	1	96.3
3	4	2	125	95.4
Overall percentage (%)	33.8	25.0	41.2	95.1

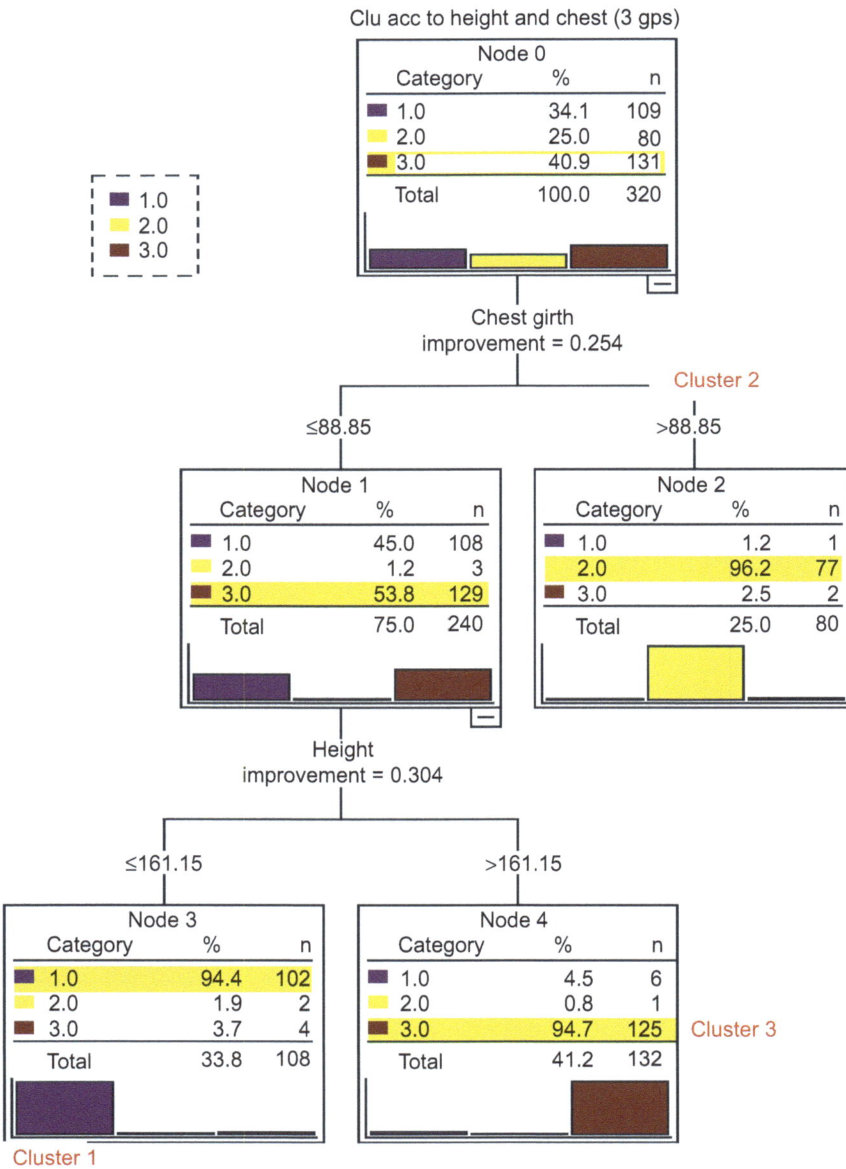

Figure 7.6 Decision tree for males (ages 13–17), upper body.

Table 7.10 **Profile of upper body types for males ages 13–17**

Node	Body type	Classified rule	*N*
1 and 3	Small	Chest girth ≤ 88.9 cm and height ≤ 161.1 cm	102
2 and 5	Medium	Chest girth ≤ 88.9 cm and height > 161.1 cm	125
6	Large	Chest girth > 88.9 cm	77

Table 7.11 **Regression tree validation, lower body (males ages 13–17)**

Observed	Predicted			
Males (13–17)	1	2	3	Percent correct (%)
1	53	2	15	75.7
2	0	135	3	97.8
3	0	5	106	95.5
Overall percentage (%)	16.6	44.5	38.9	89.7

are shown in the rules and attributes for each body type. This study implies that the body characteristics of each size vary substantially, which suggests that each rule is needed to accommodate the body variations, and in turn this provides good fit to the wearer. The same results are shown in previous studies where the coverage rules confirmed the effectiveness of sizing tables [1,2].

As can be seen, Table 7.17 shows clearly that in small size clusters the height is always less than that of the other two sizes, which means that for the tailoring or manufacturing of pants, for example, the length should be shorter as compared to the other two sizes. Note also that for small size males and females ages 7–12, the attributes are different; so again, in manufacturing pants for the lower body, as an example, for male samples height alone is significant, while for female samples both height and hip measurements needed to be considered. For the profile that Table 7.17 illustrates, height is the most important predictor variable in the younger cluster groups (7–12 years old) where height is given as the first variable used to split the sample population.

This is due to the fact that children in this age range show rapid changes in height and thus the selection of height could enhance the fit of clothing as these changes in growth are accommodated [4]. In comparison, girth variables are found to be the most significant predictor variables for both genders ages 13–17, which makes sense as the girth measurements show more distinct growth in this age group as compared to height [5]. Nevertheless, there is only one profile that shows the predictor variable is height, which is the lower body for male samples (ages 13–17).

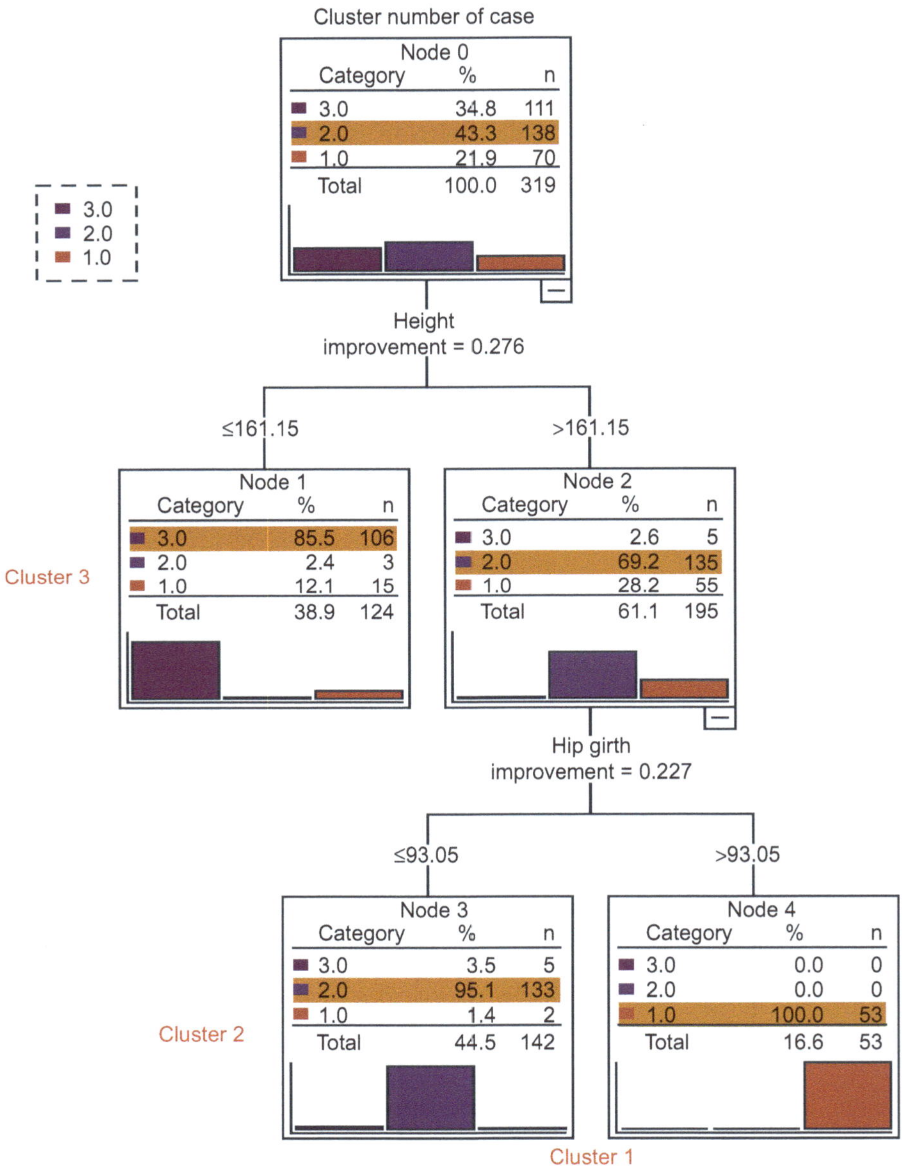

Figure 7.7 Decision tree for males (ages 13–17), lower body.

Table 7.12 Profile of lower body types for males ages 13–17

Node	Body type	Classified rule	*N*
1	Small	Height ≤ 161.1 cm	106
2 and 3	Medium	Height > 161.1 cm and hip girth ≤ 93.1 cm	135
2 and 4	Large	Height > 161.1 cm and hip girth > 93.1 cm	53

Table 7.13 Regression tree validation, upper body (females ages13–17)

Observed	Predicted			
Female (13–17)	1	2	3	Percent correct (%)
1	78	1	1	97.5
2	1	109	10	90.8
3	1	2	98	97.0
Overall percentage (%)	26.6	37.2	36.2	94.7

The findings from Table 7.16 support previous studies in which body shape was found to be a distinctive feature in teenagers, as the change in girth variables is a sign of movement toward adulthood [6,7]. These changes mean that body shape in older children (13–17) is more prominent compared to the lower age group (7–12). It is the girth variables that define this new body shape, for example, the larger bust and wider hips in teenage females [8], thus developing the sizes for this age range based on girth measurements seems appropriate for better-fit clothing.

This result is also in agreement with many previous studies in which height and girth variables were chosen as the most important predictors for body sizes [9,10,11]. Moreover, the previous cluster analyses confirmed that the division of sample group ages 7–12 is proportionate for both height and girth variables as the increment of height and girth variables show a linear relationship. This indicates that an increase in height predicts an increase in girth variables. All the significant features found in Tables 7.17 and 7.18 provide comprehensive rules for creating the size table. This table is definitely different from such tables in other countries as it is based on the local sample population, as can be seen by comparing the sizes developed in this table with those developed by previous studies of similar age ranges in other countries [12,13].

However, it is also noted that the rules of classification differ from one country to another as it is exclusively built on the said population. It can also differ when different key dimensions are used. This supports the research conducted by

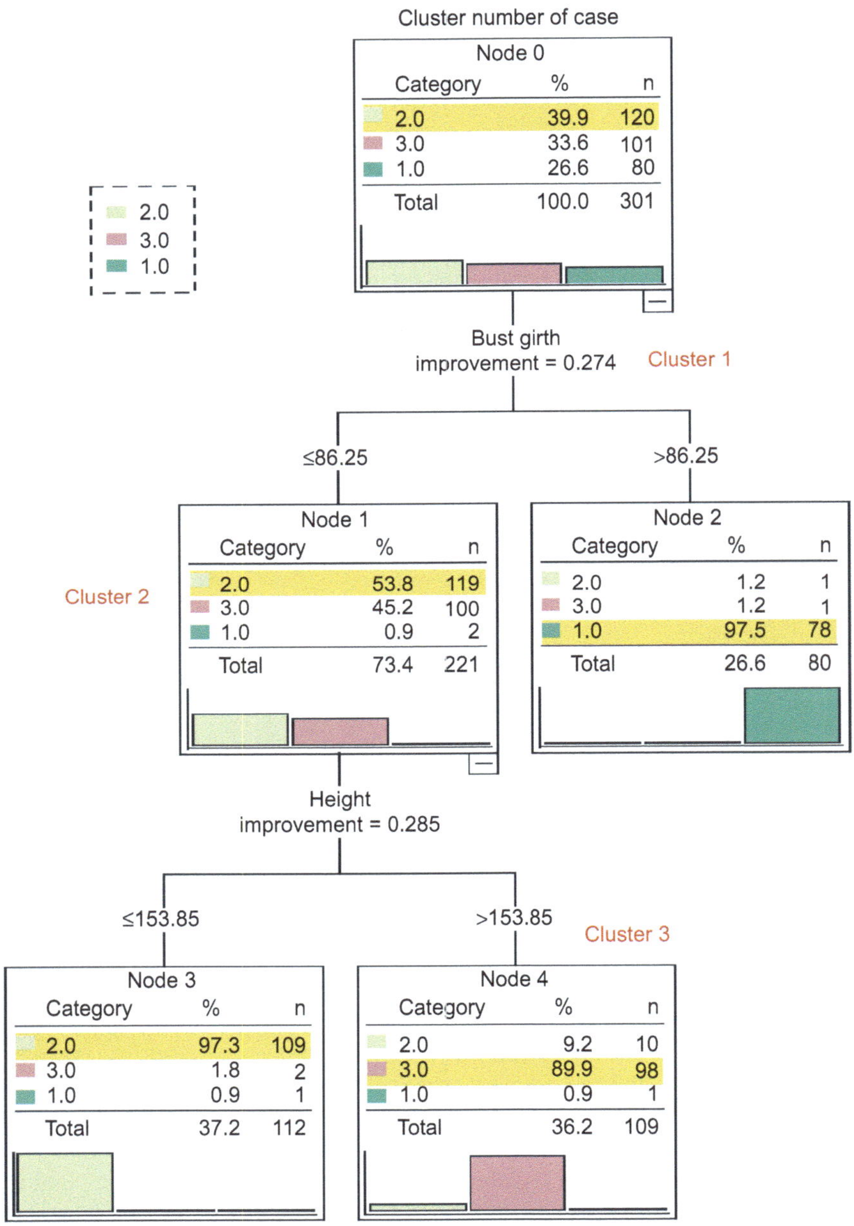

Figure 7.8 Decision tree for females (ages 13–17), upper body.

Table 7.14 **Profile of upper body types for females ages 13–17**

Node	Body type	Classified rule	*N*
1 and 3	Small	Bust girth ≤ 86.3 cm and height ≤ 153.9 cm	109
1 and 4	Medium	Bust girth ≤ 86.3 cm and height ≥ 153.9 cm	98
2	Large	Bust girth ≥ 86.3 cm	78

Table 7.15 **Regression tree validation, lower body (females ages 13–17)**

Observed	Predicted			
Female (13–17)	1	2	3	Percent correct (%)
1	98	1	3	96.1
2	18	108	0	85.7
3	0	1	63	98.4
Overall percentage	39.7	37.7	22.6	92.1

previous studies that verified that the structure of size systems differs substantially across countries and garment types and also within the same shop and same countries [14,15,16].

7.5 Conclusions

Thus, these findings add to a growing body of literature on sizing as the profiles of the sample population are confirmed to be different when compared to other countries. These newly developed profile tables, Tables 7.17 and 7.18, are therefore valuable to and practical for local manufacturers interested in providing good-fit clothing based on local anthropometric data. It can also be concluded from these results that sizing systems need to be developed based on individual countries, since differences in the human body are not the same across the globe.

As a final point, this study demonstrates clearly that all sizing systems are to be developed specifically for different populations of different countries, each according to its own anthropometric data. Even the population within the same country differs substantially, and more so across the globe, as is well supported by many previous sizing studies [2,17–21]. This finding also confirms that manufacturing of garments cannot depend on a standard grading (increasing and decreasing sizes) based on one fixed size (usually the medium size) as it is proven that the various sizes are not based on the same attributes and rules [22]. These classification rules are applied in the development of a sizing system in the next chapter.

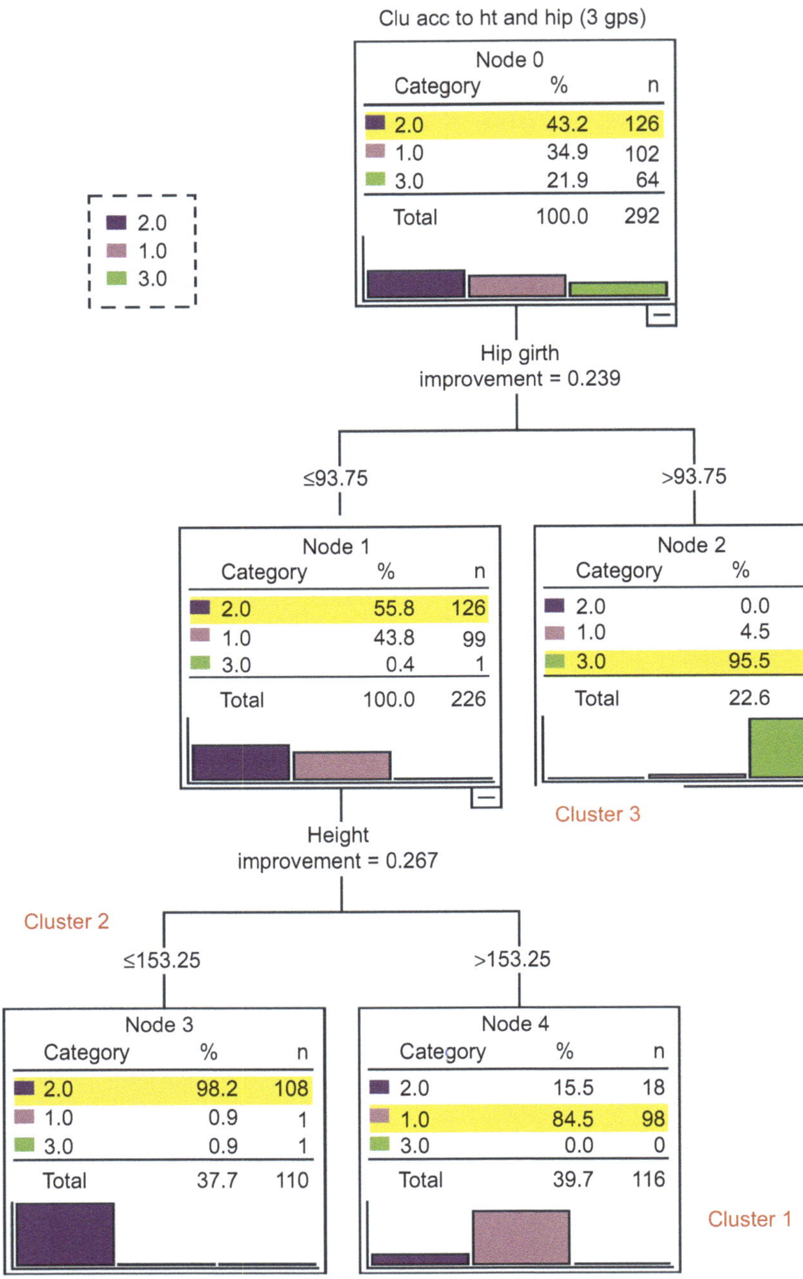

Figure 7.9 Decision tree for females (ages 13–17), lower body.

Table 7.16 Profile of lower body types for females ages 13–17

Node	Body type	Classified rule	*N*
1 and 4	Small	Hip girth ≤ 93.8 cm and height ≤ 153.3 cm	108
1 and 3	Medium	Hip girth ≤ 93.8 cm and height > 153.3 cm	98
2	Large	Hip girth > 93.8 cm	63

Table 7.17 **Body type classifications for ages 7–12**

Profile for age 7 to 12		
Body types	**Males**	**Females**
Upper body		
Small	Height < 129.6 cm	Height < 129.4 cm Bust < 65.3 cm
Medium	Height > 129.6 cm Chest < 74.6 cm	Height > 129.4 cm and < 139.7 cm Height < 129.4 cm Bust > 65.3 cm
Large	Height > 129.6 cm Chest > 74.6 cm	Height > 139.7 cm
Lower body		
Small	Height < 128.6 cm	Height < 139.7 cm Hip < 67.4 cm Height < 126.5 cm Hip > 67.4 cm
Medium	Height > 128.6 cm Hip < 76.9 cm	Height > 126.5 cm and < 139.7 cm Hip > 67.4 cm
Large	Height > 128.6 cm Hip > 76.9 cm	Height > 139.7 cm

Table 7.18 **Body type classifications for ages 13–17**

Profile for age 13 to 17		
Body types	**Males**	**Females**
Upper body		
Small	Chest < 88.9 cm Height < 161.1 cm	Bust < 86.3 cm Height < 153.9 cm
Medium	Chest < 88.9 cm Height > 161.1 cm	Bust > 86.3 cm Height > 153.9 cm
Large	Chest > 88.9 cm	Bust>86.3 cm
Lower body		
Small	Height < 161.2 cm	Hip < 93.8 cm Height < 153.3 cm
Medium	Height > 161.2 cm Chest < 93.1 cm	Hip < 93.8 cm Height > 153.3
Large	Height > 161.1 cm Chest > 93.1 cm	Hip > 93.8 cm

References

[1] H.F. Lin, C.H. Hsu, M.J. Wang, Y.C. Lin, An application of data mining technique in developing sizing system for army soldiers in Taiwan, World Sci. Eng. Acad. Soc. (WSEAS) Trans. Comput. 7 (2008) 245–252.

[2] H.L. Viktor, E. Paquet, H. Guo, Measuring to fit: virtual tailoring through cluster analysis and classification, in: Presented at 17th International European Conference on Machine Learning and the 10th European Conference on Principles and Practice of Knowledge Discovery in Database, September 18–22, Berlin, Germany, 2006.

[3] B. De Ville, Decision Trees for Business Intelligence and Data Mining: Using SAS, SAS Enterprise Miner, North Carolina, 2006.

[4] M.L. Staples, D.B. DeLury, A system for the sizing of Women's garments, Text. Res. J. 19 (1949) 346–354.

[5] R. Devi, Challenges facing the consumers in purchasing the best fit garment, in: When It Comes to Clothes, Size Matters, SIRIM Auditorium, SIRIM Berhad, December 15, 2005.

[6] R. Otieno, Approaches in researching human measurement: MMU model of utilising anthropometric data to create size charts, EuroMed J. Bus. 3 (2008) 63–82.

[7] P.R. Apeagyei, Significance of body image among UK female fashion consumers: the cult of size zero, the skinny trend, Int. J. Fash. Des. Technol. Edu. 1 (2008) 3–11.

[8] The International Association of Athletics Federations (IAAF), Introduction to coaching theory: growth and development, in: IAAF Coaching Theory Manual, 2009.

[9] Y. Kang, H.S. Choi, W.H. Do, A study of the apparel sizing of children's wears. An analysis of the size increments utilized in children's wear based on an anthropometric survey, J. Korean Home Econ. Assoc. Engl. Ed. 2 (2001).

[10] A.S. Ariadurai, T.P.G. Nilusha, M.R. Dissanayake, An anthropometric study on Sri Lankan school children for developing clothing sizes, J. Soc. Sci. 19 (2009) 51–56.

[11] K. Yeosun, C. Hei-Sun, H.D. Woei, A study of the apparel sizing of children's wear – an analysis of the size increments utilized in children's wear based on an anthropometric survey, J. Korean Home Econ. Assoc. Engl. Ed. 2 (2001) 96–110.

[12] M.J. Chung, H.F. Lin, J.J. Mao, J. Wang, The development of sizing systems for Taiwanese elementary-and high-school students, Int. J. Ind. Ergon. 37 (2007) 707–716.

[13] C.Y. Jongsuk, C.R. Jasper, Garment-sizing charts: an international comparison, Int. J. Cloth. Sci. Technol. 5 (1993) 28–37.

[14] D. Ujević, B.B. Šajatović, R. Hrženjak, K. Doležal, I. Karabegović, V. Petrovečki, Contribution to anthropometric surveys and principal comparisons of the garment size system, in: Presented at the 86th Textile Institute World Conference, Fashion and Textiles: Heading towards New Horizons, Hong Kong, China, November 18–21, 2008.

[15] D. Ujevi, I. Karabegovi, Anthropometry and the comparison of garment size systems in some European countries, Coll. Anthropol. 29 (1) (2005) 71–78.

[16] R. Otieno, C. Harrow, G. Lea-Greenwood, The unhappy shopper, a retail experience: exploring fashion, fit and affordability, Int. J. Retail Distrib. Manag. 33 (4) (2005) 298–309.

[17] E. Bye, K.L. LaBat, M.R. Delong, Analysis of body measurement systems for apparel, Cloth. Text. Res. J. 4 (2) (2006) 66–79.

[18] R. Otieno, C. Fairhurst, The development of new clothing size charts for female Kenyan children. Part I: using anthropometric data to create size charts, J. Text. Inst. 1754–2340 91 (2000) 143–152.

[19] M.J. Chung, M.J. Wang, The development of sizing systems for school students, in: In 36th International Conference on Computers and Industrial Engineering. Taipei, Taiwan, 2006.

[20] D. Gupta, A mathematical approach for developing a garment sizing system for Indian population, in: International Conference on High Performance Textiles & Apparels (HPTEX 2004), 2004.
[21] S. Ashdown, S. Loker, Improved Apparel Sizing: Fit and Anthropometric 3D Scan, 2004. National Textile Center Annual Report: Nov 2004.
[22] D. Gupta, B. Gangadhar, A statistical model for developing body size charts for garments, Int. J. Cloth. Sci. Technol. 16 (2004) 458–469.

Developing apparel sizing systems for children and teenagers

8

8.1 Introduction: developing sizing systems for children and teenagers

In this chapter, the sizing system for each group is presented and discussed. There are two steps involved in developing the sizing system as shown in Fig. 8.1 based on Fig. 3.1 in Chapter 3. The first step is to develop the sizing system and the second step is to validate the sizing system for better fit. Section 8.2 discusses the importance of body shapes in sizing systems. In Section 8.3, the sizing system development is completed for the four different sample groups according to the classifying rules (Tables 7.15 and 7.16) from Chapter 7. In Section 8.4, based on Step 9, the size tables for both children and teenagers are presented. Section 8.5, based on Step 10 presents the validation table for entire samples, which are the solutions for selecting better fit sizes. The proposed sizing systems are reported and discussed for the following four samples: female (7–12 years old), male (7–12 years old), female (13–17 years old), and male (13–17 years old). The size table is developed according to upper and lower body garments as suggested by the ISO standard [1,2]. The sizing system is developed by separating the samples into the correct size category (small, medium, and large), which uses the CRT classification method of which the IF/THEN statements are employed on the samples using the profile tables (Tables 7.15 and 7.16).

8.2 The importance of body sizes and shapes in sizing system for children

Based on the previous chapters, it can be seen that the body data of the children are being classified according to different groups systematically. The data mining process analyzes the data comprehensively and segregates the children data into well-defined sizes and shapes. As can be seen from the data mining process from Chapters 6 and 7, an exhaustive statistical process was utilized to mine the collective body dimensions so that a good sizing system could be developed based on the classified groups that gives very structured body classifications. Interesting body shape variations were captured from the data set to show that children belong to different subsizes, which determine good fit of clothing.

The objective of the children sizing system is to create sizes for a particular group of children. The idea is to ensure that the sizes of clothes are dictated by two important

Clothing for Children and Teenagers. http://dx.doi.org/10.1016/B978-0-08-100226-1.00008-4
Copyright © 2016 Elsevier Ltd. All rights reserved.

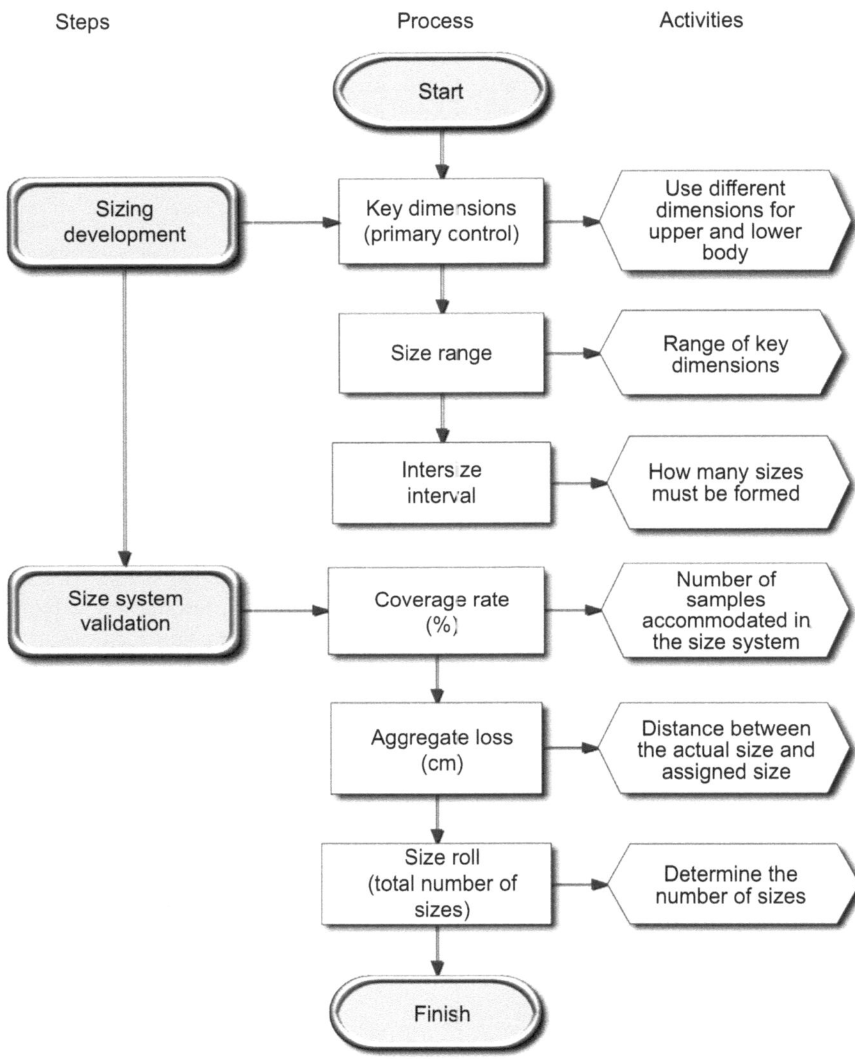

Figure 8.1 Diagram of sizing process.

body components, which are the sizes and shapes of human beings. Furthermore, children are built in different proportions, which means that they can be different sizes (small, medium, or large) and shapes (slender, petite, stout, medium, or big), which have components of body dimensions that are not proportionate. In addition, a sizing system created for children can also be more challenging as the growth patterns of children are not stable. This is due to the fact that children grow rapidly from age 7 until 17, which is the coverage age for this research.

Children of the age range 7–17 years old show rapid growth. Therefore, the analysis of shapes and sizes is more critical and necessary as the range of body dimensions is bigger in children as compared to adult body data. When the range of body dimensions is bigger, this gives more varied sizes and shapes. Thus, careful decisions need to be made to develop the sizing system that needs to accommodate most of the sample population.

The sizing system will also reflect the body proportions of the children, which will be meaningful for different sizes as they are not based on standard increments but instead cater to the nature of individual body proportions based on the varied sizes and shapes. According to the research conducted by Glock and Kunz [3], a sizing system will also include a range of sizes based on gradations of dimensions for a body type.

A good sizing system that is based on sizes and shapes also plays an important psychological role for children's dressing where the sizes developed will enhance the fit for each child. Children are already becoming wise consumers today whereby they are aware of their actual body size [4]. Therefore, wearing the right sizes seems to be the right move to flatter their physical look and also so they are not judged by their peers for bad fit. It is noted that children by the age of 12 have already accepted their own body and understand the body image needed to find the right sizes of clothes [5]. Girls and boys who are going through hormonal changes—the teenagers—seem to be more sensitive toward their body image such as shapes, sizes, and proportions [6]. This called for another objective of developing a good sizing system that is a basis for providing the selection of finding the right sizes for different children.

Clothes are a major item that connects children to their peers and lets them fit in. Thus, children and teenagers are susceptible to negative reactions from peers if they do not wear the right size clothes [7]. Clothes that looked unfit will make them looked untidy and this may create undesirable perceptions by their peers. Children today also understand the importance of well-fitted garments on the body based on their sizes and shapes [8]. Therefore, a sizing system that can reflect body types and shapes of current populations is very important to accommodate the particular fit preferences [9].

Accordingly, this means that wearing good-fitting clothes is not only about knowing your size or measurements but also about knowing your body shape [10]. The body is three-dimensional, so when the patterns are developed, they need to consider the three-dimensional aspects. From this research, it was found that children can have different variations of body type, shapes, and proportions. Some are tall with smaller girth dimensions. Some are tall with large girth dimensions, and some are tall with medium girth dimensions. On the other hand, some are short with small girth dimensions, which is petite size, etc. In addition, some children have different body proportions where they have smaller size dimensions in the upper body and bigger size dimensions in the lower body. This lead to an understanding that children do not grade linearly and body proportions are one of the components and combinations needed to be considered for a good sizing system to be developed for children.

8.3 Sizing system development

The proposed sizing systems are reported and discussed for the following four samples: female (7–12 years old), male (7–12 years old), female (13–17 years old), and male (13–17 years old). The size table is developed according to upper and lower body garments as suggested by the ISO 3636:1977 and ISO 3636:1977. The total numbers

Table 8.1 **Final output of sizing tables**

Sample groups	Upper body garment	Lower body garment	Total size table
Females: ages 7–12	3 size tables	3 size tables	6 size tables
Males: ages 7–12	3 size tables	3 size tables	6 size tables
Females: ages 13–17	3 size tables	3 size tables	6 size tables
Males: age 13–17	3 size tables	3 size tables	6 size tables
	12 size tables	12 size tables	**24 size tables**

of sizing systems are depicted in Table 8.1 in which 24 sizing tables are developed for the four samples.

The development of a sizing system for this study is based on the range and size interval as mentioned in Chapter 3. Then, to separate the samples into the correct size category (small, medium, and large) the IF/THEN statements are employed on the samples using the profile tables (Tables 7.15 and 7.16).

8.3.1 Key dimensions, size interval and range

Deciding the size range can be quite challenging, as the aim of any sizing system is to accommodate as many people in the sample size but with the right number of size rolls. The size intervals are determined for both the key dimensions before the samples are classified according to the correct sizes.

Hence, after examining the range of the key dimensions, size interval for height is 8 cm for ages 7–12 and 6 cm for ages 13–17. The different height size interval for different age groups is because height range found in ages 7–12 is bigger as compared to ages 13–17. Therefore, for large height range, a bigger size interval is used and vice versa for lower size range. For example: height range for 7–12 is 40 cm so using 8 cm will give five sizes, whereas for ages 13–17 is 30 cm; using 6 cm also gives five sizes. In all, there are five height sizes for all sample groups.

In comparison, size interval for the girth is either 5 or 6 cm in order to get a maximum of seven sizes. For example, if girth range is 35 cm, 5-cm interval is used to get exactly seven sizes, whereas if the girth range is 36 cm, 6-cm interval is used to get six sizes. Hence, the results show that there are either five, six, or seven sizes for girth key dimensions for each sample group, respectively, in which the details are shown in the next sections. According to Hsu et al. [11] in his studies, he confirmed the use of 4–6 cm as the girth size interval. In addition, the height interval is 8 and 4 cm for other dimensions when they created size systems for children. Moreover, in developing a children size system, Otieno uses 6 cm for height interval and 2 cm for hip and chest girth [12]. Most countries report using 4–6 cm for the girth interval as it is accepted and confirmed by the experts to cover the most people as possible with the fewest sizes [13–16].

Consequently, this finding shows that the total number of sizes for each sample is estimated to be less than 36 sizes for each sample group covering either the upper or lower body in which there are either 25, 30, or 35 sizes. For example, height has five sizes and girth has five, six, or seven sizes. Hence, the minimum total size = 5 × 5 = 25 sizes or maximum total sizes = 5 × 7 = 35 sizes. These ranges are found to give a reasonable estimated size roll (total number of sizes) to accommodate the entire range found in each sample group. However, the size roll mentioned is just an estimation to cover the range for each sample. Caution must be applied, as not the entire assigned size roll matches the sample body sizes. Therefore, the size roll can be less or retained as an estimate in order to accommodate the entire sample. Only after application of classifying rules will the total of size rolls be identified.

As a result, each size system is presented on a bivariate plot using the key dimensions. For each sizing system, the vertical axis is presented by height and the horizontal axis is presented by girth variables depending on the upper or lower body. This bivariate plot includes the size intervals that determine the size role for each sample size. In addition, this graph visually depicts the distribution of the three body types, which fulfilled the classification rules.

8.3.2 Male sizing system (7–12 years old) for upper body garments

Fig. 8.2 depicts the scatter plot of sample size according to the key dimensions of height and chest girth. The height range is divided into five subgroups using an interval of 8 cm, which resulted in five sizes: 114, 122, 130, 138, and 146 cm in order to cover the height range of 114–154 cm. As for chest representation, a set of seven sizes are designed at 55, 60, 65, 70, 75, 80, and 85 cm. The size interval for both height and chest girth indicated there are 35 sizes to accommodate the entire sample population.

After the assigned sizes are established, the classifying rules are applied using the IF/THEN statements in order to classify them into respective sizing classes: small,

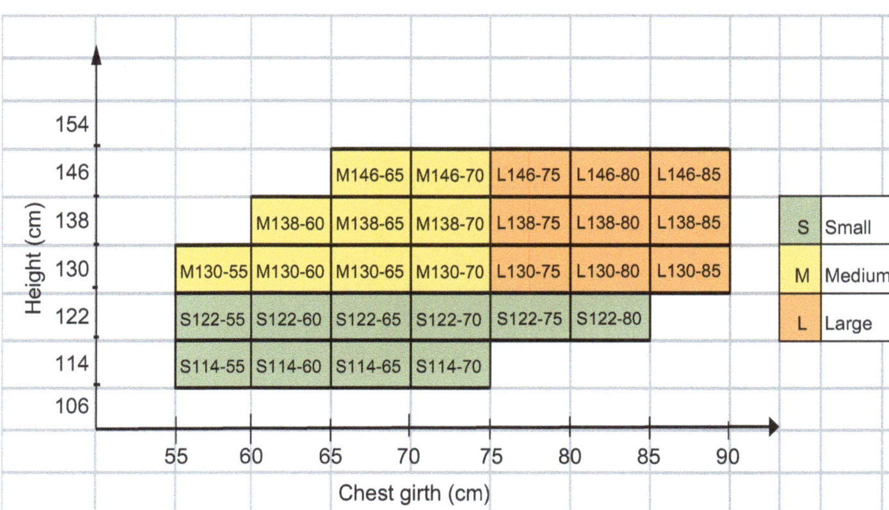

Figure 8.2 Distribution graph of height versus chest girth for upper body, males (7–12 years old).

medium, and large. An examination of the classifying rules indicated that 10 sizes are suitable for small-sized samples, nine sizes for medium-sized and large-sized samples, respectively. Overall, 28 sizes are developed for this sample size. It can be seen from Fig. 8.2 that seven sizes are not included in the size classes. Among them are the sizes at the far end. For example, small sizes S114-75 or S114-85 and so forth. This result indicates that assigned sizes may not be able to cater to the real size as it is based on range. The details of coverage for each size are presented in the next section.

8.3.3 Male sizing system (7–12 years old) for lower body garments

Fig. 8.3 shows the distribution of sizes according to lower body type for males 7–12 years old. The size interval for height is 8 cm with the same sizes mentioned previously for height range of 114–151 cm. The hip girth has size range of 55–89.9 cm with size interval of six; 56, 62, 68, 74, 80, and 86 cm. There are six sizes for hip girth and five sizes for height, estimating 30 sizes for this sample group.

Using the classifying rules, nine sizes are developed for small-sized children, six sizes for medium-sized children, and nine sizes for large-sized children. Overall, 24 sizes are developed for this sample group instead of 30 estimated sizes. This indicates that the assigned range is excellent to accommodate the samples, resulting in fewer sizes as recommended by previous research [17,18].

8.3.4 Female sizing system (7–12 years old) for upper body garments

Fig. 8.4 shows the distribution of sizes according to the body types for females between 7 and 12 years old. These three body types are produced by cluster analysis technique

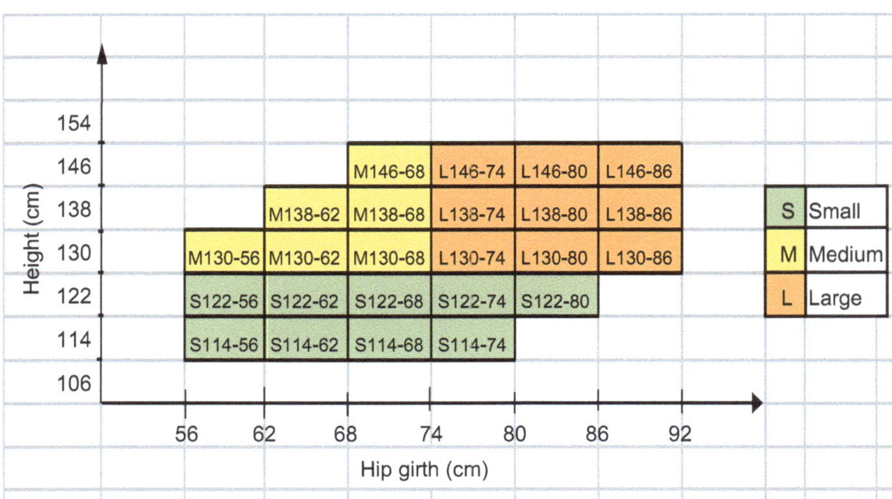

Figure 8.3 Distribution graph of height versus hip girth for lower body, males (7–12 years old).

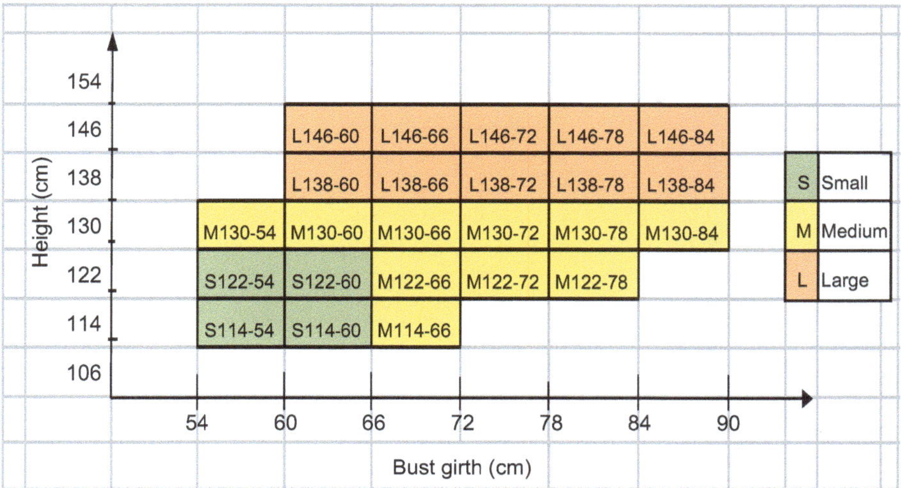

Figure 8.4 Distribution graph of bust girth versus height for upper body, females (7–12 years old).

and verified by decision tree analysis in Chapter 5. The control variables for upper body are height and bust girth. The size interval for height is 8 cm, and for bust girth is 6 cm. The range for height in this population is 114–153.7 cm. This range is divided into six subgroups using an interval of 8 cm. The size interval was chosen based on the best coverage for these samples; in addition, ISO/TC133 commonly uses that as a sizing interval [19]. For the bust, the total range is 31.5 cm (56–87.5 cm); this range was divided into six subgroups using an interval of six: 54, 60, 66, 72, 78, and 84 cm. The distribution graph of the key dimensions is illustrated in Fig. 6.3, with bust girth as the x-axis and height as the y-axis.

From Fig. 8.4, it can be seen that this sample population is divided into three sizes: small, medium, and large, resulting in 24 size rolls.

For the small size category, four sizes are provided with height 114–129.9 cm and 54–65.9 cm bust girth measurement. For the medium size category, 10 sizes are provided with variations of height of 114–137.9 cm with bust measurements of 54–89.9 cm. Ten sizes are provided for the last group (large size) for tall children measuring 138–153.9 cm with bust measurement of 60–89.9 cm.

8.3.5 *Female sizing system (7–12 years old) for lower body garments*

In Fig. 8.5, for the lower body, the control variables are height and hip girth. The range for height is 114–153.7 cm and the range for hip girth is 58–97 cm. The size interval for height is 8 cm and for the hip girth are 5 and 6 cm. For small size samples, the size interval for hip girth is 5 cm. For medium and large size, the size interval is 6 cm. These two different size intervals are applied in order to get minimum sizes (seven) as compared to eight sizes if using a 5-cm interval. It is assumed that using seven sizes is

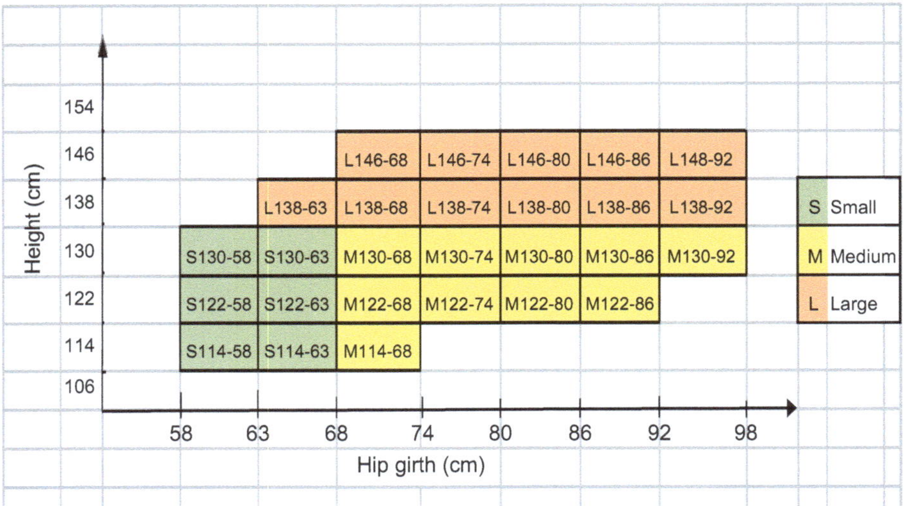

Figure 8.5 Distribution graph of height versus hip girth for lower body, females (7–12 years old).

much better than using eight sizes. This assumption was proven correct as estimation of 35 sizes resulted in only 31 sizes to accommodate the sample size. In contrast, if using 5 cm for girth intervals, it produces 40 sizes and perhaps the accommodation would not be so accurate.

As given by the classifying rules (Table 7.15), the samples are classified according to the correct sizes as shown in Fig. 8.5. For small size category, seven sizes are found to accommodate samples with height of 114–137.9 cm and with 58–73.9 cm hip girth measurement. For the medium size category, there are nine sizes for children 122–137.9 cm tall with hip measurements of 68–97.9 cm. There are 11 sizes for large size for samples 138–153.9 cm tall with bust measurement of 63–97.9 cm.

8.3.6 *Male sizing system (13–17 years old) for upper body garments*

Fig. 8.6 shows the distribution of sizes according to body type for male samples ages 13–17. The control variables for upper body are height and chest girth. The size interval for height is 6 cm and for chest, girth is 6 cm. The range for height in this population is 146–176 cm, which gives six subgroups using an interval of five: 146, 152, 158, 164, and 170 cm.

For the chest, the total range is 70–106 cm; this range was divided into 5-cm intervals of seven sizes: 70, 76, 82, 88, 94, 100, and 106 cm. Hence, 35 sizes are assigned to male samples. However, after assigning the samples to each size, only 26 sizes match the different sizes of male samples. This indicates that fewer sizes are achieved (26 sizes) in comparison to the estimated sizes (35 sizes) in order to accommodate the sample size. The figures of samples in each size are calculated in the size validation section.

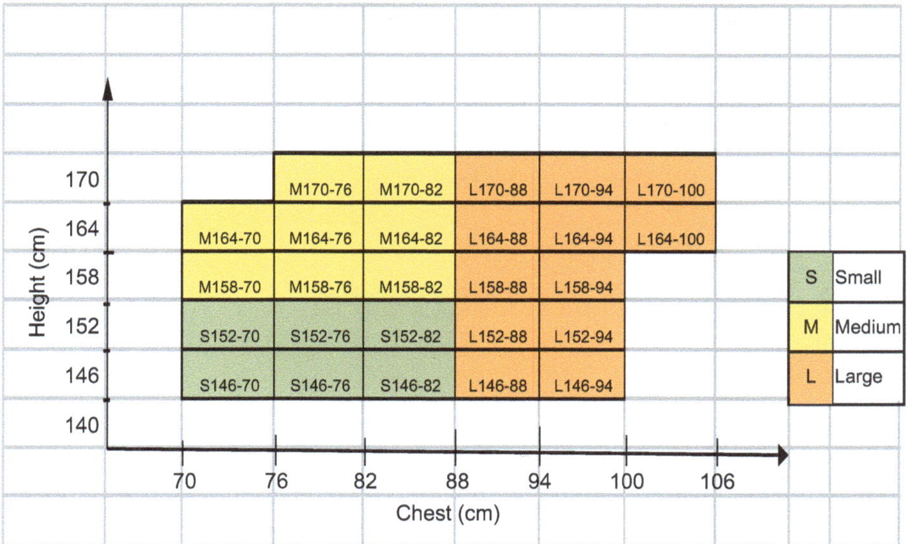

Figure 8.6 Distribution graph of height versus chest girth for upper body, males (13–17 years old).

Fig. 8.6 also depicts the distribution of samples according to the different body types as revealed in the decision tree analysis. There are six sizes designed for small body type as prescribed by the classifying rules. In addition, eight sizes are classified into medium sizes while another eight sizes are set for large sizes, totaling up to 26 sizes for this sample.

8.3.7 Male sizing system (13–17 years old) for lower body garments

The distribution graph for the lower body sizing system is given in Fig. 6.6. The size interval for height is 6 cm with five sizes: 146, 152, 158, 164, and 170 cm, whereas hip girth has a size interval of 5 cm. The sizes can be seen from the *x*-axis in Fig. 6.6. The following graph depicts eight sizes for small body types, 10 for medium size samples, and 11 for large size, for a total of 29 sizes as compared to 35 sizes estimated for this sample population. This proved that the size interval selected for this sample group is accurate in accommodating the entire samples as the sizes are found to be smaller than the estimated size. From Fig. 8, there are 11 sizes suitable for small body type, 10 sizes for medium body type, and another eight sizes for large body type.

8.3.8 Female sizing system (13–17 years old) for upper body garments

For teenage females (13–17 years old), the control variables for upper body are height and bust girth as seen in Fig. 8.8. The size interval for height is 8 cm and for bust, the

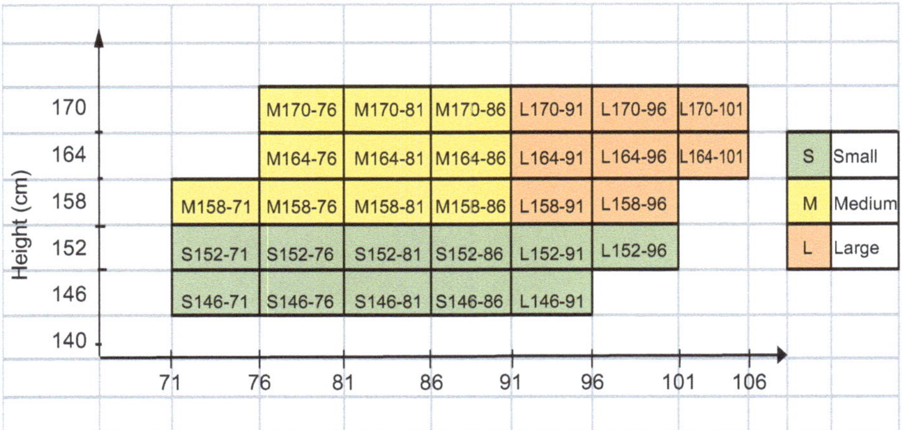

Figure 8.7 Distribution graph of hip girth versus height for lower body, males (13–17 years old).

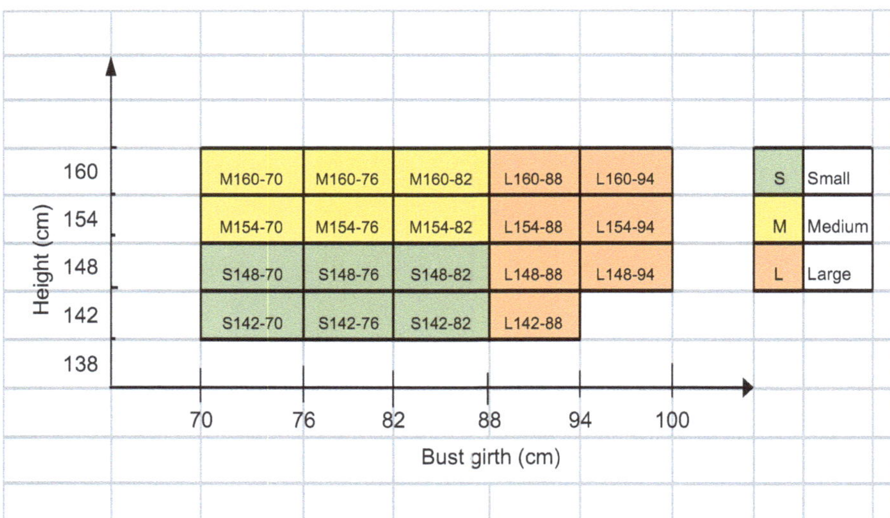

Figure 8.8 Distribution graph of height versus bust girth for upper body, females (13–17 years old).

girth is 6 cm. The range for height in this population is 142–166 cm, which is the same for upper and lower body. The range is divided into five subgroups using an interval of 6 cm, which are 142, 148, 154, 160, and 166 cm. For the bust, the range is 70–98 cm; this range is divided using an interval of six: 70, 76, 82, 88, 94, and 100 cm. With the size interval, the height gives five subgroups whereas the hip gives six subgroups totaling 30 sizes designed to accommodate the female samples (ages 13–17). Fig. 8.8 indicates the outcome of the body type selection based on the classifying rules. In total, there are 19 sizes suitable for the sample population. Six sizes are developed for small size samples, six sizes for medium size samples, and another seven sizes for large size samples.

There are 30 sizes designed to accommodate this entire sample size but the outcome indicates that only 19 sizes are able to cover the whole samples, which means that the samples are found to be in one of these 19 sizes. This result proves that the size interval is accurate and so does the rule that dictates which samples get into what sizes based on the IF/THEN statement. However, the coverage of the samples is not known yet as it is calculated in the next section. Smaller size rolls can indicate the size system is good only if the coverage of the samples is high above 80% as compared to smaller size rolls and smaller percentage of coverage.

8.3.9 Female sizing system (13–17 years old) for lower body garments

Fig. 8.9 depicts the distribution of female samples (ages 13–17) according to lower body. As can be seen from the figure, the height size intervals are the same as in the upper body of the same sample population whereas the hip girth shows a range of 75–105 cm. This gives a size interval of 5 cm with six sizes: 75, 80, 85, 90, 95, and 100 cm. In total, 30 sizes are developed to accommodate the samples in which only 22 sizes are required to accommodate the whole samples. From Fig. 8.9, there are eight sizes suitable for small sizes, seven sizes for medium, and another seven sizes for large body type.

8.4 Problems and solutions in selecting sizes for better fit

The main problem in any children ready-to-wear (RTW) clothing industry is to provide the sizes for the majority of the children with a variety of shapes and sizes. This is especially true due to the rapid body growth of the children ages 7–17. It is also the

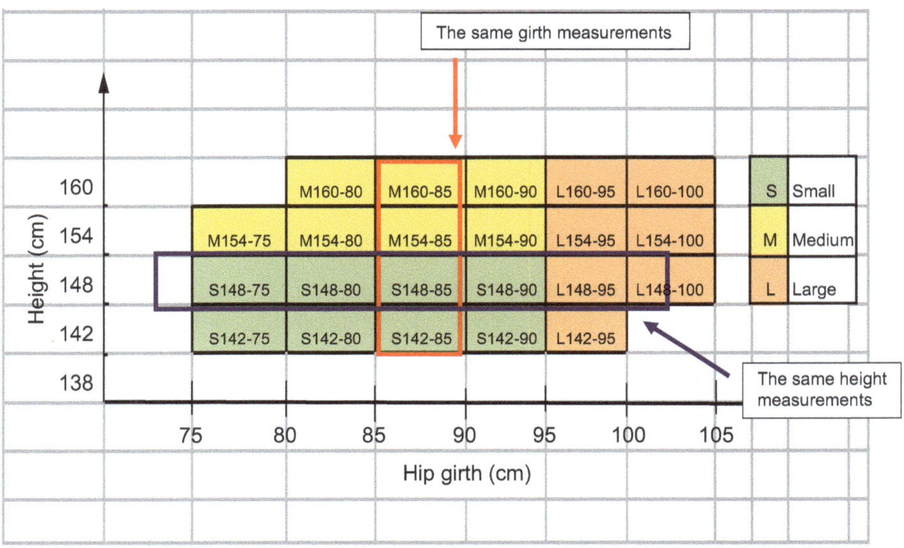

Figure 8.9 Distribution graph of hip girth versus height for lower body, females (13–17 years old).

fact that all RTW sizes are made up from size ranges that need to accommodate the majority sizes of these children. It is also noted that some of the sizing systems that denote the sizes of clothing were developed based on the linear proportionate body shapes and sizes. This means that for all the different sizes, the increments of sizes are based on systematic increments. Based on this research, it can be seen from Figs. 8.1–8.9 that all the sizes are created when the data is analyzed statistically to get the real proportionate body based on the data measurements. There is also evidence that with more data about the real children in a population, this will give more realistic sizes to give a good fit. The main solution for sizing development is to ensure that the sizes are based on the right body sizes and shapes according to realistic proportions.

Following the development of the sizing table, the solution to better fit is to validate the size system according to three criteria: size roll (total number of sizes), accommodation rate (%), and aggregate loss (cm). The validation of a sizing system using this method has been recognized in many previous studies as it statistically confirms the effectiveness of the sizing system [20–23].

It is noted from this study that the challenge lies in finding the right size interval in order to accommodate as many samples as possible in one sizing system. It is therefore the goal of any sizing system to find the best size interval to signify the best coverage of the target population while having not so many size rolls. From the overall results, the entire samples are highly accommodated with appropriate numbers of size rolls, which proves the size intervals selected are correct. However, it is also noted in some of the sizing systems that there are some sizes that are not densely accommodated.

The efficiency and accuracy of each size table for every sample group is shown in the next section. For each section, there are two tables: one discusses the overall result of the size table, and the other discusses the details of the size tables that designate the acquired sizes for each cluster obtained from the size population. Both contain the elements of the size table. The elements presented are size roll, key sizes, accommodation rate, and aggregate loss for each body type. The tables also review the sizes that are not densely distributed (those that have less than 2% coverage). The 2% coverage as mentioned in Kwona et al.'s [24] study showed low accommodation rate, while in other previous studies less than 1% coverage is considered as sparsely accommodated [11,25]. This means that the coverage rate regarded as loosely accommodated is below 2%, and therefore in this study all the sizes below the 2% rate are marked with asterisk (*).

8.4.1 Male size validation (ages 7–12) for upper body garments

As shown in Table 8.2, 10 sizes are created for the small size, which consists of 235 samples, whereas nine sizes are created for the medium size, which consists of 208 samples. Nine more sizes are developed for large size, which consists of 115 samples. Therefore, a total of 28 sizes are constructed to accommodate 558 samples, with coverage of 99.1%. This suggests that majority of the samples measured match the newly assigned clothing sizes. In addition, the aggregate loss for the assigned sizes is 3.0, which is lower than the ideal value (3.6), indicating a good result.

Table 8.3 shows the distribution of samples and aggregate loss for each size created. As can be seen from the table, some sizes contain many samples and some only

Table 8.2 **Size validation for upper body, male (ages 7–12)**

Body type	Size roll	n (%)	Mean aggregate loss (cm)
Small	10	235 (41.8)	3.0
Medium	9	208 (36.9)	3.0
Large	9	115 (20.4)	3.0
Total	28	558 (99.1)	3.0

Table 8.3 **Size distribution for upper body, male (ages 7–12)**

Size roll	Body types	Key dimensions (cm)		Accommodation rate		Aggregate loss (cm)
		Height	Chest girth	n	%	
1	Small	114–121.9	55–59.9	56	9.9	3.1
2			60–64.9	33	5.9	3.1
3			65–69.9	7	1.2*	3.0
4			70–74.9	3	0.5*	3.1
5		122–129.9	55–59.9	32	5.7	3.1
6			60–64.9	58	10.3	3.0
7			65–69.9	29	5.2	2.9
8			70–74.9	7	1.2*	3.0
9			75–79.9	7	1.2*	2.9
10			80–84.9	3	0.5*	3.1
11	Medium	130–137.9	55–59.9	11	2.0	3.1
12			60–64.9	42	7.5	3.1
13			65–69.9	42	7.5	3.0
14			70–74.9	24	4.3	3.1
15		138–145.9	60–64.9	14	2.5	3.1
16			65–69.9	38	6.7	3.1
17			70–74.9	18	3.2	3.1
18		146–151.9	65–69.9	8	1.4*	2.8
19			70–74.9	11	2.0	2.9
20	Large	130–137.9	75–79.9	19	3.4	3.0
21			80–84.9	13	2.3	3.1
22			85–89.9	6	1.1*	3.1
23		138–145.9	75–79.9	21	3.7	3.0
24			80–84.9	21	3.7	3.1
25			85–89.9	9	1.6*	3.1
26		146–153.9	75–79.9	9	1.6*	2.8
27			80–84.9	10	1.8*	2.9
28			85–89.9	7	1.2*	2.9
TOTAL				**558**	**99.1**	**3.0**

a few, like the ones marked with an asterisk. Nevertheless, out of 35 sizes established from the key dimensions, only 28 sizes are needed to match all the samples belonging to this group. The aggregate loss is also shown for each size and reveals lower values (2.8–3.1) than the ideal value of 3.6, indicating that the assigned sizes are close to the actual sizes.

It is also noted in Table 8.3 that there are 11 sizes, in which five sizes are for small body type, one size for medium body type, and five sizes for large body type. These sizes show less than 2% covered, which means it has only a few samples included in the size. This means that the sizes may not be considered for mass production as too few samples in a size have cost implications and this is not economically effective. Thus, if these sizes are removed, the total sizes retained are 20 with 90% coverage. Ninety percent coverage confirms the efficiency of the size table, as only 10% of the samples are not included in the sizes.

8.4.2 Male size validation (ages 7–12) for lower body garments

When developing sizes for the lower body, the results yield nine sizes for small body type, six sizes for medium body type, and another nine sizes for large body type, totaling 26 sizes for the entire group; 40.6% of the samples are covered in small size, compared to 28.7% for the medium size, and 30% for the large size. This is almost a 99% result as 99.3% of the samples fall into one of the three body types. The aggregate loss values are below the ideal value although they are higher than in the upper body. However, the value shows that the aggregate loss is good (Table 8.4).

Table 8.5 shows all the values (size roll, number of samples covered, and aggregate loss) for each body type. As can be seen, there are only four sizes that show a coverage of less than 2%, which overall is an excellent result as it means most of the sizes created are efficient.

8.4.3 Female size validation (ages 7–12) for upper body garments

From Table 8.6 it can be seen that this sample population falls into three sizes: small with 34.7% of the sample population, medium with 33.3% of the sample population, and large with 31.2% of sample population, totaling 99.3% coverage for the entire

Table 8.4 Size validation for lower body, male (ages 7–12)

Body type	Size roll	n (%)	Mean aggregate loss (cm)
Small	9	221 (40.6)	3.0
Medium	6	156 (28.7)	3.0
Large	9	163 (30.0)	2.9
Total	26	540 (99.3)	3.2

Table 8.5 Size distribution for lower body, male (ages 7–12)

Size roll	Body types	Key dimensions (cm)		Accommodation rate		Aggregate loss (cm)
		Height	Hip girth	n	%	
1	Small	114–121.9	56–61.9	53	9.7	3.0
2			62–67.9	22	4.0	2.9
3			68–73.9	11	2.0	3.0
4			74–79.9	3	0.6*	3.0
5		122–129.9	56–61.9	40	7.4	3.0
6			62–67.9	50	9.2	2.9
7			68–73.9	24	4.4	3.0
8			74–79.9	14	2.6	3.0
9			80–85.9	4	0.7*	2.9
10	Medium	130–137.9	56–61.9	10	1.8*	3.0
11			62–67.9	47	8.6	2.9
12			68–71.9	37	6.8	3.0
13		138–145.9	62–67.9	17	3.1	2.9
14			68–73.9	31	5.7	3.0
15		146–153.9	68–73.9	14	2.6	3.0
16	Large	130–137.9	74–79.9	28	5.1	3.0
17			80–85.9	22	4.0	3.0
18			86–91.9	12	2.2	2.9
19		138–145.9	74–79.9	27	5.0	3.0
20			80–85.9	28	5.1	3.0
21			86–91.9	16	2.9	2.8
22		146–153.9	74–79.9	8	1.5*	2.9
23			80–85.9	12	2.2	2.9
24			86–91.9	10	1.8*	2.8
TOTAL				**540**	**99.3**	**3.2**

Table 8.6 Size validation for upper body, female (ages 7–12)

Body type	Size roll	n (%)	Mean aggregate loss (cm)
Small	4	182 (34.7)	2.6
Medium	10	175 (33.3)	2.7
Large	10	164 (31.2)	2.6
Total	24	521 (99.3)	2.6

group. This indicates that the sizes developed successfully accommodate almost all the samples.

The 24 sizes obtained from these samples were categorized as small with four sizes, and medium and large with 10 sizes each. The average aggregate loss calculated for each body type is lower than in the male samples (7–12), indicating that the sizes assigned fit the measurements of this sample population better as compared with the sizes assigned for males of the same age range.

Table 8.7 shows that more samples were grouped under the medium and large size to match the variation of body size and shape found in these samples. Specifically, the large and medium size samples are tall and average height with varied bust measurements, while for the small sizes the bust measurements show much less variation. This indicates that for small size where the majority are at age 7, the growth of bust is slow and therefore as they reach older ages the bust blossom results in more variations of bust girth for height.

Overall, from Table 8.7, the aggregate loss is good for these samples as the range is 2.5–3.0cm. In addition, there are only six sizes out of 24 sizes showing loose accommodation as coverage (<2%). This is shown as four in medium size and two in the large size. However, the overall result of accommodation is high and thus this system is considered efficient, as almost 100% of the samples are accommodated although with some less densely accommodated.

Table 8.7 Size distribution for upper body, female (ages 7–12)

Size roll	Body types	Key dimensions (cm)		Accommodation rate		Aggregate loss (cm)
		Height	Bust girth	n	%	
1	Small	114–121.9	54–59.9	54	10.3	2.6
2			60–65.9	14	2.7	2.5
3		122–129.9	54–59.9	60	11.4	2.6
4			60–65.9	54	10.3	2.5
5	Medium	114–121.9	66–71.9	8	1.5*	2.6
6		122–129.9	66–71.9	29	5.5	2.5
7			72–77.9	11	2.1	2.7
8			78–83.9	4	0.8*	2.5
9		130–137.9	54–59.9	13	2.5	2.7
10			60–65.9	46	8.8	2.7
11			66–71.9	31	5.9	2.6
12			72–77.9	23	4.4	2.8
13			78–83.9	7	1.3*	2.6
14			84–89.9	3	0.6*	3.0
15	Large	138–145.9	60–65.9	15	2.9	2.5
16			66–71.9	36	6.9	2.5
17			72–77.9	16	3.0	2.6
18			78–83.9	17	3.2	2.5
19			84–89.9	6	1.1*	2.7
20		146–153.9	60–65.9	3	0.6*	2.6
21			66–71.9	17	3.2	2.6
22			72–77.9	18	3.4	2.7
23			78–83.9	22	4.2	2.5
24			84–89.9	14	2.7	2.8
TOTAL				**521**	**99.3**	**2.6**

8.4.4 Female size validation (ages 7–12) for lower body garments

In this sample population for lower body, 36.4% are small size, 33.3% are medium size, and 30.4% are large size. The sizes developed accommodate 100% of the total sample size, an excellent result. This means that the sizes assigned match all of the samples' actual body measurements. The size roll consists of seven sizes for small body type, nine sizes for medium, and 11 for large. The total number of sizes is 27. The aggregate loss of goodness of fit for each body type is good, and the mean of aggregate loss for this entire sample is lower than the ideal value, which is 2.6 cm. Thus, the recommended size tables are quite accurate for two reasons: (1) they accommodate all the samples and (2) the aggregate loss value is smaller than the ideal value (Table 8.8).

Table 8.9 shows each size with the number of samples, accommodation rate, and aggregate loss. The large group contains the fewest samples but the highest number of sizes as compared to the small group, which has the most samples but the lowest number of sizes. It can be seen that the large size samples reflect more variations in body shape and size as compared to other sample groups.

Overall, there are nine sizes with less than 2% coverage, as seen in the size table. Two of the sizes came from the small group, four from the medium group, and three from the large group. The aggregate loss ranges from 2.5 to 2.8 cm, lower than the ideal value, which indicates that the sizes proposed for this sample group are accurate. This proposed size table is efficient, as the coverage is 100%, meaning that all samples were accommodated by the assigned sizes. However, three assigned sizes contain relatively few samples (less than 1% coverage), which may mean that they are not so practical and economical for mass production. Therefore, for the sizing system, if the nine sizes were omitted, the remainder of 21 sizes still demonstrates high coverage rate (95%).

8.4.5 Male size validation (ages 13–17) for upper body garments

As shown in Table 8.10, the small group (22.2%) was accommodated by six sizes, the medium (48.3%) by eight sizes, and the large group (25.4%) by 12 sizes. Thus, the sizes created successfully accommodate 95.9% of the sample population. For each body type, the aggregate loss of goodness of fit is well within acceptable parameters, and the average aggregate loss of 2.9 cm proves the size table is quite accurate.

Table 8.8 Size validation for lower body, female (ages 7–12)

Body type	Size roll	n (%)	Mean aggregate loss (cm)
Small	7	187 (36.4)	2.7
Medium	9	171 (33.3)	2.7
Large	11	156 (30.4)	2.6
Total	27	514 (100)	2.6

Table 8.9 **Size distribution for lower body, female (ages 7–12)**

Size roll	Body types	Key dimensions (cm)		Accommodation rate		Aggregate loss (cm)
		Height	Hip girth	n	%	
1	Small	114–121.9	58–62.9	40	7.8	2.7
2			63–67.9	25	4.9	2.7
3			68–73.9	8	1.6*	2.8
4		122–129.9	58–62.9	32	6.2	2.6
5			63–67.9	57	11.1	2.7
6		130–137.9	58–62.9	4	0.8*	2.6
7			63–67.9	21	4.1	2.6
8	Medium	122–129.9	68–73.9	33	6.4	2.8
9			74–79.9	29	5.5	2.6
10			80–85.9	8	1.6*	2.8
11			86–91.9	2	0.6*	2.6
12		130–137.9	68–73.9	34	6.6	2.7
13			74–79.9	37	7.2	2.5
14			80–85.9	16	3.1	2.8
15			86–91.9	8	1.6*	2.6
16			92–97.9	4	0.8*	2.6
17	Large	138–145.9	63–67.9	6	1.2*	2.6
18			68–73.9	16	3.1	2.7
19			74–79.9	27	5.3	2.5
20			80–85.9	17	3.3	2.7
21			86–91.9	15	2.9	2.5
22			92–97.9	7	1.4*	2.5
23		146–153.9	68–73.9	5	1.0*	2.7
24			74–79.9	12	2.3	2.5
25			80–85.9	15	2.9	2.8
26			86–91.9	21	4.1	2.5
27			92–97.9	15	2.9	2.5
Total				**514**	**100.0**	**2.6**

Table 8.10 **Size validation for upper body, male (ages 13–17)**

Body type	Size roll	n (%)	Mean aggregate loss (cm)
Small	6	72 (22.2)	2.9
Medium	8	150 (48.3)	2.9
Large	12	81 (25.4)	3.0
Total	26	303 (95.9)	2.9

Table 8.11 depicts the elements of the size table. The large body type has 12 sizes, more than the other two body types although it consists of the fewest number of samples. There are many variations in the large group, indicating inefficiency, as the number of each size contains only few samples. This will affect the economy of the

Table 8.11 **Size distribution for upper body, male (ages 13–17)**

Size roll	Body types	Key dimensions (cm)		Accommodation rate (cm)		Aggregate loss (cm)
		Chest girth	Height	n	%	
1	Small	70–75.9	146–151.9	13	4.1	2.9
2			152–157.9	16	5.1	2.7
3		76–81.9	146–151.9	11	3.5	3.0
4			152–157.9	23	7.3	2.9
5		82–87.9	146–151.9	4	1.3*	3.0
6			152–157.9	5	1.6	2.9
7	Medium	70–75.9	158–163.9	18	5.7	2.9
8			164–169.9	4	1.3*	2.9
9		76–81.9	158–163.9	21	6.6	3.0
10			164–169.9	32	10.1	3.0
11			170–175.9	10	3.2	2.9
12		82–87.9	158–163.9	16	5.1	3.0
13			164–169.9	27	8.5	3.0
14			170–175.9	22	7.0	2.9
15		88–93.9	146–151.9	4	1.3*	3.0
16	Large		152–157.9	3	0.9*	2.9
17			158–163.9	6	1.9	3.0
18			164–169.9	20	6.3	3.0
19			170–175.9	9	2.8	2.9
20		94–99.9	146–151.9	4	1.3*	3.1
21			152–157.9	3	0.9*	2.9
22			158–163.9	5	1.6	3.0
23			164–169.9	11	3.5	3.0
24			170–175.9	8	2.5	2.9
25		100–105.9	164–169.9	4	1.3*	3.1
26			170–175.9	4	1.3*	3.0
TOTAL				**303**	**95.9**	**2.9**

producer as compared to the medium size, for example, only eight sizes accommodate the 150 samples in the medium body type compared to the 12 sizes needed to accommodate the 81 samples in large body type.

Table 8.11 also shows that the large body type has six sizes with less than 2% coverage and that in all, eight sizes have less than 2% coverage. Hence, if these sizes are removed from the size table, it retains a total of 18 sizes with 87% coverage. This result indicates that, if the size system has less densely accommodated sizes and these sizes were eliminated, then the efficiency of the table is reduced. However, in this study all the sizes were retained as the selection of sizes are up to the manufacturers to decide. Moreover, the same sizes less than 1% coverage rates were retained in the sizing system in other previous study [25].

Table 8.12 **Size validation for lower body, male (ages 13–17)**

Body type	Size roll	n (%)	Mean aggregate loss (cm)
Small	11	84 (26.6)	3.0
Medium	10	155 (49.1)	2.9
Large	8	54 (17.5)	3.0
Total	29	293 (93.2)	2.9

8.4.6 Male size validation (13–17 years old) for lower body garments

Table 8.12 tabulates the size elements for this group for lower body. The small group shows 26.6% accommodation, the medium group 49.1%, and the large group 17.5%. Hence, 29 sizes were proposed for this sample with a total coverage of 93.2%. This result confirms the efficiency of the size table, as the sizes are able to accommodate a majority of the samples. Moreover, the aggregate loss of fit shows good results for each body type (3.0, 2.9, and 3.0 cm, respectively) amounting to an average of 3.0 cm for the whole group.

Table 8.13 depicts the size elements of the lower body for male samples ages 13–17. It can be seen from the table that there are six sizes in the small body type for which the coverage is less than 2%, one size in the medium, and four in the large. This result indicates that although the coverage is high, some sizes are not densely packed. Again, it is left to the manufacturers to choose which sizes are best suited to their target population.

8.4.7 Female size validation (13–17 years old) for upper body garments

Table 8.14 proposes 19 sizes to accommodate the entire sample population for this group: six each for small and medium body types and seven for the large body type. Almost 40% of the samples were accommodated in the small body type, 37.5% in the medium, and 21.9% in the large. The results show 99.3% coverage for this sample group. In addition, the aggregate loss for each body type is excellent as they show much lower values than the ideal, and the average aggregate loss of fit is excellent at 2.5 cm. Details of each size are presented in Table 6.14.

Table 8.15 shows an efficient size table as most of the samples were assigned to one of the proposed sizes. The efficiency of the table is confirmed high as only two sizes were less than 2% coverage. In addition, the aggregate loss for the entire size table shows an excellent value of 2.5 cm, well below the ideal value, meaning that the proposed sizes are accurate for this group.

8.4.8 Female size validation (13–17 years old) for lower body garments

As can be seen from Table 8.16, there are eight sizes for the small body type and seven each for the medium and large body type. In total, 22 sizes were developed to

Table 8.13 **Size distribution for lower body, male (ages 13–17)**

Size roll	Body types	Key dimensions (cm)		Accommodation rate (cm)		Aggregate loss (cm)
		Height	Hip girth	n	%	
1	Small	146–151.9	71–75.9	10	3.2	3.0
2			76–80.9	10	3.2	3.1
3			81–85.9	6	1.9*	3.0
4			86–90.9	5	1.6*	2.9
5			91–95.9	3	0.9*	3.0
6		152–157.9	71–75.9	11	3.5	3.0
7			76–80.9	19	6.0	3.1
8			81–85.9	12	3.8	2.9
9			86–90.9	2	0.6*	2.8
10			91–95.9	3	0.9*	3.0
11			96–100.9	3	0.9*	3.0
12	Medium	158–163.9	71–75.9	8	2.5	2.9
13			76–80.9	21	6.6	2.9
14			81–85.9	18	5.7	2.8
15			86–90.9	10	3.2	2.7
16		164–169.9	76–80.9	15	4.7	3.1
17			81–85.9	30	9.5	3.0
18			86–90.9	22	7.0	2.9
19		170–175.9	76–80.9	4	1.3*	3.0
20			81–85.9	8	2.5	2.9
21			86–90.9	19	6.0	2.8
22	Large	158–163.9	91–95.9	4	1.3*	2.9
23			96–100.9	2	1.0*	3.0
24		164–169.9	91–95.9	12	3.8	3.0
25			96–100.9	11	3.5	3.0
26			101–105.9	4	1.3*	3.1
27		170–175.9	91–95.9	9	2.8	2.9
28			96–100.9	7	2.2	3.0
29			101–105.9	5	1.6*	3.0
TOTAL				**293**	**93.1**	**2.9**

Table 8.14 **Size validation for upper body, female (ages 13–17)**

Body type	Size group	n (%)	Mean aggregate loss (cm)
Small	6	120 (39.9)	2.6
Medium	6	113 (37.5)	2.7
Large	7	66 (21.9)	2.6
Total	19	299 (99.3)	2.6

Table 8.15 **Size distribution for upper body, female (ages 13–17)**

Size roll	Body type	Key dimensions (cm)		Accommodation rate		Aggregate loss (cm)
		Bust girth	Height	n	%	
1	Small	70–75.9	142–147.9	12	4.0	2.7
2			148–153.9	32	10.6	2.7
3		76–81.9	142–147.9	10	3.3	2.6
4			148–153.9	45	15.0	2.6
5		82–87.9	142–147.9	7	2.3	2.5
6			148–153.9	14	4.7	2.6
7	Medium	70–75.9	154–159.9	14	4.7	2.9
8			160–165.9	5	1.7*	2.8
9		76–81.9	154–159.9	38	12.6	2.7
10			160–165.9	23	7.6	2.6
11		82–87.9	154–159.9	18	6.0	2.7
12			160–165.9	15	5.0	2.6
13	Large	88–93.9	142–147.9	2	0.7*	2.5
14			148–153.9	10	3.3	2.5
15			154–159.9	21	7.0	2.7
16			160–165.9	6	2.0	2.6
17		94–99.9	148–153.9	8	2.7	2.6
18			154–159.9	13	4.3	2.8
19			160–165.9	6	2.0	2.7
TOTAL				**299**	**99.3**	**2.05**

Table 8.16 **Size validation for lower body, female (ages 13–17)**

Body type	Size roll	n (%)	Mean aggregate loss (cm)
Small	8	122 (41.8)	2.6
Medium	7	112 (38.4)	2.7
Large	7	53 (18.2)	2.7
Total	22	287 (98.5)	2.7

accommodate this sample group; 41.8% of the samples were accommodated in the small body type, 38.4% in the medium type, and 18.2% in the large type, for 98.5% coverage for this sample group. This indicates that the size table is efficient as the coverage shows a very high percentage. In Table 8.17, the coverage was shown for each size and the efficiency is viewed for every size available. The nonefficient sizes are marked in the sizing system to show if they were loosely accommodated. The aggregate loss for each body type is lower than the ideal value, which confirms that the sizes are quite accurate. In total, the average aggregate loss of fit is 2.5 cm. Table 8.16 views the overall result and the details of each size are presented in Table 8.17.

Table 8.17 **Size distribution for lower body, female (ages 13–17)**

Size roll	Body types	Key dimensions (cm)		Accommodation rate		Aggregate loss (cm)
		Hip girth	**Height**	**n**	**%**	
1	Small	75–79.9	142–147.9	5	1.7*	2.6
2			148–153.9	6	2.1	2.8
3		80–84.9	142–147.9	9	3.1	2.5
4			148–153.9	35	12.0	2.7
5		85–89.9	142–147.9	8	2.7	2.6
6			148–153.9	39	13.4	2.8
7		90–94.9	142–147.9	4	1.4*	2.5
8			148–153.9	16	5.5	2.7
9	Medium	75–79.9	154–159.9	4	1.4*	2.8
10		80–84.9	154–159.9	22	7.5	2.7
11			160–165.9	6	2.1	2.6
12		85–89.9	154–159.9	22	7.5	2.8
13			160–165.9	22	7.5	2.7
14		90–94.9	154–159.9	21	7.2	2.7
15			160–165.9	15	5.1	2.6
16	Large	95–99.9	142–147.9	3	1.0*	2.6
17			148–153.9	5	1.7*	2.8
18			154–159.9	18	6.2	2.8
19			160–165.9	5	1.7*	2.7
20		100–104.9	148–153.9	6	2.1	2.7
21			154–159.9	11	3.8	2.7
22			160–165.9	5	1.7*	2.7
TOTAL				**287**	**98.3**	**2.7**

The sizes created for this sample group for lower body were tabulated in Table 8.17. Overall, the same number of sizes were created for each body type with one extra in the small body type, which contains more samples. Overall, the sizes are efficient with most of the sizes showing coverage greater than 2% except for seven sizes, which is seven out of 21 sizes in total.

For each size, the aggregate loss of fit is between 2.5 and 2.8. Since the value shown is less than the ideal value, this confirms the accuracy of each size proposed for the samples in this sample group. For better viewing of Table 8.17, this space is left blank as to show the table in the next page.

8.5 Size validation according to gender and age

Several striking results emerge from this section for each sample group. The first is that applying the classifying rules from decision tree analysis resulted in unequal numbers of sizes in each cluster group. These cluster groups reflect the body types for each sample group (small, medium, and large). However, these unequal numbers of sizes in each cluster group are able to produce efficient proposed sizes, as shown by the high percentage of total coverage for the sample groups. Thus, this study has proven that unequal numbers of sizes in cluster groups have no adverse effect on building a good sizing system, in contrast to the earlier study, which found that equal numbers of sizes in cluster groups provided the basis for a good new standard sizing system [26]. In addition, the results also demonstrated that the aggregate loss of fit for each cluster group (body type) is accurate and that the accuracy of the total size table is good, as is seen from the aggregate loss value, which is lower than the ideal value.

Second, it is obvious from each table that the set sizes of height and girth variables for each sample group are similar, for example, Table 8.1, sizes S114-65, S114-60, S122-60, M122-65, L146-65, L146-70, etc. However, although these sizes for height and girth variables are similar for each individual sample group, the resulting sizes are distinctly different. This is because the classifying rules were applied and proved to have different rules for each body type, which resulted in different sizes for different body types.

This is also the reason why the smallest size and the largest size from each cluster group differ significantly. This result is different from previous studies, which showed that the extreme sizes in each cluster group were similar when the cluster analysis technique is applied. Consequently, different classification techniques prove to give different results. Nevertheless, the most important criteria for a good sizing system are distinctive size range, efficiency, and accuracy. It is therefore proven from this research that using classifying rules produces distinctive sizes and an efficient and accurate size table.

Third, it is clearly seen that individuals in each body type can be well fitted even though using a standard product. The reason for this is that the size ranges are continuous, so no gaps are found from one size to another size. Table 8.3 shows, for example, that small size 1 height is 114–121.9 and chest girth is 55–59.9 cm, while for small size 2, height is 114–121.9 cm and chest girth is 60–64.9 cm. All chest girths from 55 to 64.9 are covered, therefore, a person with height of 118 cm and chest of 58 cm is in size 1 while a person with height of 122 cm and chest of 63 cm is in size 2. If height is 121 cm and chest is 61, the person can choose either size 1 or size 2. If size 1 is chosen, then the clothing is a good fit. On the other hand if size 2 is chosen, there is extra space for growth.

As mentioned before, one significant result was shown in the size validation table where all the coverage rates are distinctly high. However, when the distribution tables were examined, it was suggested that within the same size table, some sizes cover the samples tightly and some sizes cover them loosely. This can be seen, for example, in Table 8.13 where size 7 covers only 0.6% of the sample population while size 17 covers up to 9.5% of the same population. This indicates that in any one sample group, some sizes accommodate the samples densely and some sizes do not.

From previous literature, the preset percentage of the lowest accommodation rate for each size ranges from 0.1% to 2% [24,25]. Therefore, in each distribution size table all the sizes with less than 2% coverage are marked with an asterisk (*). Tables 8.14 and 8.15, for example, show two different efficiency results. Table 8.13 has eight sizes with less than 2% coverage and therefore resulted in 93.1% coverage as a whole, while Table 8.15 has only one size with less than 2% coverage, resulting in 99.3% coverage as a whole.

However, in this study the lowest percentage accommodation is found to be 0.5%, which is not as low as the ISO standard minimum in which sizes as low as 0.1% are included in the size table [24]. It is also noted that those sizes with less than 2% coverage rate are the ones that did not show 100% cluster results, which are mainly the large size samples with characteristics of large girths with short to tall height. Therefore, it is decided that all sizes shall be retained for the outcome results as some of the sizes with low percentage coverage are still useful for customized tailoring, as they contain only a few samples.

8.6 Conclusions

To conclude this chapter, a summary of Table 8.18 is given to show the validation chart for the entire sizing system built for the different sample populations in this research. The sample involves children ages 7–17, which are classified into two main groups: (1) children (7–12 years old) and (2) teenagers (13–17 years old). As shown in Table 8.18, it is significant that sizing systems for both genders ages 7–12 successfully accommodated 99% of the sample population measured and have minimal total sizes of 24–28. The aggregate loss is 2.6–3.0 cm (specifically 3.0, 3.0, 2.6, and 2.7 cm, respectively), well below the ideal value of 3.6 cm (obtained in Chapter 3). Hence, the goodness of fit for this system is considered accurate and good. Moreover, the efficiency of the size system proves to be high, as the average coverage factor is above 90%.

On the other hand, for samples 13–17 years old, the sizing systems for both genders successfully accommodated 93–99% of the sample population with minimal total sizes of 20–28. The aggregate loss value is 2.5–2.9 cm, less than the ideal value, which proves the accuracy of the sizes for fit evaluation.

Having classified the population into three groups according to height and girth variables, eight sizing systems are created for this sample population with 200 sizes in order to accommodate the wide range of variations that exist in children ages 7–17. Furthermore, as can be seen from Table 8.3, for each sample group, the number of sizes is below 30 and the coverage is excellent at 93–99% of accommodation rate.

In sum, this research has successfully developed an efficient sizing system with high coverage of population, accurate sizes based on good aggregate loss of fit, and a reasonable size roll that is lower than other size project results [21,26]. For similar research, the author is comparing the results to two children's sizing research projects conducted in Taiwan and Korea. Both of them have similar age ranges. In comparison, the size system covering Taiwanese children ages 6–18 developed 12 sizing systems with less than 36 sizes in each sample group [21]. In total, 315 sizes were developed to accommodate the entire sample size in which the coverage rate ranged from 85.4% to 91.2%. The aggregate loss ranged from 5.23 to 6.91 cm; this is higher than the ideal value of 5.08 cm, which uses four key dimensions to cluster the sample groups. The Taiwan sizing system used an interval of 8 cm for height and 4 cm for all girth

Table 8.18 **Validation table for sizing system for both genders and age groups**

Age group	**7–12 years old**				**13–17 years old**			
Gender	**Male**		**Female**		**Male**		**Female**	
Garment category	**Upper**	**Lower**	**Upper**	**Lower**	**Upper**	**Lower**	**Upper**	**Lower**
Key dimensions	Height & chest girth	Height & hip girth	Height & bust girth	Height & hip girth	Height & chest girth	Height & hip girth	Height & bust girth	Height & hip girth
Coverage (%)	99	99	99	99	96	93	99	98
Aggregate loss (cm)	3.0	3.2	2.6	2.6	2.9	2.9	2.5	2.7
Number of sizes (size roll)	28	24	24	27	27	28	20	22

variables as compared to this study, which uses 8 and 6 cm for height, 5 and 6 cm for girth variables. Thus, the Taiwan system sizes are more in quantity due to the size interval of girth variables being smaller and the size range wider than in this research.

In addition, the Korean sizing system covering children 6–13 years old developed 43 garment sizes with a 4-cm size interval for chest girth with 5 cm for height. The Korean sizes seemed to have the most number of sizes as compared to both this research and the Taiwan sizing system for a similar age range [27].

It is also important to consider different market segments for a new standard sizing system. For instance, in order to enhance the sizing system, the same research could be replicated using different populations such as adult females and males, plus sizes, and special needs consumers.

This study used the manual method of measuring the body for the collection of anthropometric data. However, with the advancement of technology many studies in sizing research have moved toward collecting the data using 3D body scanners. Future research could adopt the same technique using technology to facilitate data collection and reduce the time involved.

Furthermore, the usage of different analysis techniques such as artificial neural network (AI) and other advanced computer programming techniques could be explored in order to produce the sizing system that can give different size results and efficiency.

References

[1] ISO 3636, Size Designation of Clothes – Men's and Boys' Outerwear Garments, 1977.

[2] ISO 3637, Size Designation of Clothes – Women's and Girls' Outerwear Garments, 1977.

[3] R.E. Glock, G. Kunz, Apparel Manufacturing Sewn Product Analysis, fourth ed., Person Education, Upper saddle River, New Jersey, 2005.

[4] O.J. Panff, H.K. Hyllegard, Y. Ruoh-Nan, An investigation of mothers' and tween daughters' clothing preferences and purchase intentions toward a prosocial clothing company, J. Fash. Market. Manag. 18 (1) (2014).

[5] P.B. Makhanya, H.M. de Klerk, K. Adamski, A. Mastamet Mason, Ethnicity, body shape differences and female consumer's apparel fit problems, Int. J. Consum. Stud. 38 (2) (2014) 183–191.

[6] K.J. Isaksen, S. Roper, The commodification of self-esteem: branding and British teenagers, Psychol. Market. 29 (3) (2012) 117–135.

[7] E. Gentina, Understanding the effects of adolescent girl's social positions within peer groups on exchange practices, J. Consum. Behav. 13 (1) (2014) 73–80.

[8] O.J. Panff, H.K. Hyllegard, Y. Ruoh-Nan, M.A. Littrell, Mother and teen daughter socialization toward ethical apparel consumption, Fam. Consum. Sci. Res. J. 43 (10) (2014).

[9] T. Tselepis, H.M. de Klerk, Early adolescent girls' expectations about the fit of clothes: a conceptual framework, SSN 0378 5254, J. Fam. Ecol. Consum. Sci. 32 (2004).

[10] L.D. Romeo, Y.A. Lee, Exploring current ready-to-wear apparel attributes for plus-size female teens, Fam. Consum. Sci. Res. J. 43 (2) (2014) 145–149.

[11] H.C. Hsu, H.F. Lin, M.J. Wang, Developing female size charts for facilitating garment production by using data mining, J. Chin. Inst. Ind. Eng. 24 (2007) 245–251.

[12] R. Otieno, C. Fairhurst, The development of new clothing size charts for female Kenyan children. Part I: using anthropometric data to create size charts, J. Text. Inst. 91 (2000) 143–152, 1754–2340.

[13] A.S. Ariadurai, T.P.G. Nilusha, M.R. Dissanayake, An anthropometric study on Sri Lankan school children for developing clothing sizes, J. Soc. Sci. 19 (2009) 51–56.
[14] J.M. Winks, Clothing Sizes: International Standardization, 1997.
[15] ISO/TR 10652, Standard Sizing Systems for Clothes, 1991.
[16] J.C. Yoon, C.R. Jasper, Garment-sizing systems: an international comparison, Int. J. Cloth. Sci. Technol. 5 (1993) 28–37.
[17] A. Petrova, Creating sizing systems, in: in: S.P. Ashdown (Ed.), Sizing in Clothing Developing Effective Sizing Systems for Ready-to-Wear Clothing, 2007, pp. 75–82. Woodhead Publishing, Cambridge.
[18] M.-È. Faust, The Use of Standard Sizes in Women's Ready to Wear and Consumer Fit, Unpublished Doctoral Dissertation, École Polytechnique de Montréal, Montreal, Quebec, 2007.
[19] ISO 3637, Size Designation of Clothes-Women's and Girls' Outerwear Garments, 1977.
[20] S.L. Amirul Ihsan, Beauty: one size doesnt fit all, in: New Strait Times, April 27th. Kuala Lumpur, 2009.
[21] M.J. Chung, H.F. Lin, J.J. Mao, J. Wang, The development of sizing systems for Taiwanese elementary-and high-school students, Int. J. Ind. Ergonom. 37 (2007) 707–716.
[22] D. Gupta, B.R. Gangadhar, A statistical model for developing body size charts for garments, Int. J. Cloth. Sci. Technol. 16 (2004) 458–469.
[23] M.J. Chung, M.J. Wang, The development of sizing systems for school students, in: 36th International Conference on Computers and Industrial Engineering. Taipei, Taiwan, 2006.
[24] O. Kwona, K. Jungb, H. Youb, H.-E. Kimc, Determination of key dimensions for a glove sizing system by analyzing the relationships between hand dimensions, Appl. Ergonom. 40 (2009) 762–766.
[25] ISO/TR 10652, Standard Sizing Systems for Clothes, 1991.
[26] R. Bagherzadeh, M. Latifi, A.R. Faramarzi, Employing a three-stage data mining procedure to develop sizing system, World Appl. Sci. J. 8 (8) (2010) 923–929.
[27] K. K. 0050–1999, Sizing System for Men's and Boy's Garments, Korean Standards Association, Seoul, 1999.

Designation and validation of apparel sizes for children and teenagers

9

9.1 Introduction: size designation and validation

This is the final step in the sizing system development where the size designation is provided for better selection of fit (Fig. 9.1). Section 9.2 discusses the importance of size designation and Mondoform labeling. In Section 9.3, the validation of sizing designation and number of sizes is described for both gender and age group. In Section 9.4, the female size designation tables (ages 13–17) are given. Section 9.5 discussed the children sizing designation potential application to different groups of children and teenagers. The last two sections include the conclusion and recommendations for future work.

9.2 The importance of size designation—Mondoform labeling

As mentioned in previous literature, a good sizing system must be able to accommodate as many people as possible [2–4]. Most importantly, good sizing systems are those that integrate anthropometric data of the targeted population. The size labels transmit size information using numeric codes, which simplify the task of finding a good-fit garment for consumers. Past literature has shown that size labels must be accurate and based on information about the sizing development. Faust et al. [5] added that size labels were originally designed to provide information that should help consumers select their correct size. Thus, this indicates that size designation must be well developed, which defines the label correctly to ensure that customers buy according to the sizes and the problem of fit can be minimized.

Size designation is defined as the coding system that allows the wearer to identify the body dimensions for which a particular garment is intended based on the size categories. Size categories are sizes that are grouped together taking into consideration the variability (gender, age, and body shapes) of the targeted consumers [6]. Size designations can be represented many different ways including numbers (8,10,12,14), letters (S,M,L,XL), age (7–8 years old, 9–10 years old), or numerical codes (eg, S114-55, M121-70, L154-80) giving the relevant dimensions such as height and chest. Methods of size designation have changed over the course of the development of the clothing industry and differ from one country to another.

Clothing for Children and Teenagers. http://dx.doi.org/10.1016/B978-0-08-100226-1.00009-6
Copyright © 2016 Elsevier Ltd. All rights reserved.

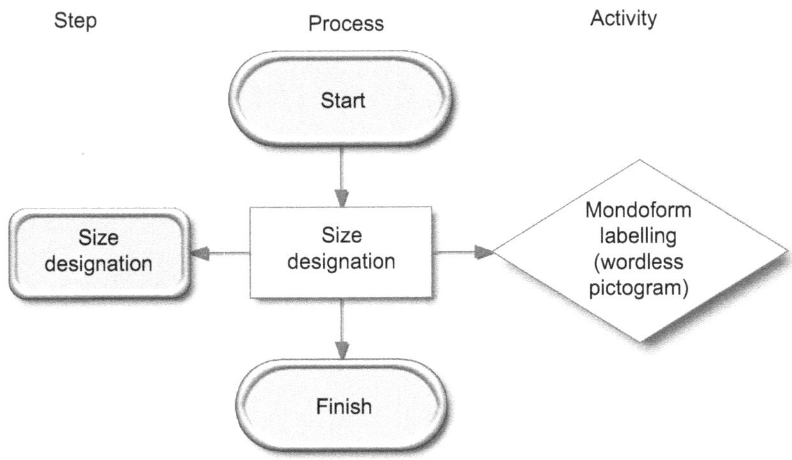

Figure 9.1 Size designation chart.

Another example would be garment sizes that are given in numbers (eg, 6), in age (eg, 9–11 years old), or letters (eg, S for small). Very few size categories are labeled according to numerical codes listing the body dimensions used except for infants, toddlers, and children between the ages of 0 months and 6 years. For girls' and boys' sizes from 7 to 12 years old, the common way of labeling garment size is by age. For teenage girls and boys ages 13–19 years old, size labels are usually letters or numbers. However, these size labels can vary tremendously from one manufacturer to another, which creates confusion and difficulties for consumers [4].

Size labels usually contain information that helps consumers choose clothing in the retail shop. In addition, size labels are put on ready-to-wear (RTW) clothes to show the suggested sizes that will fit the consumers according to their body dimensions [7]. However, Workman and Lentz [8] claimed that most size labels have no information about the body measurements on which the sized are based. There is only the numeric or alphabetical size designation [8].

This results in confusion among consumers because when consumers shop for garments, they first attempt to determine whether a garment will fit by looking at the size label. If they cannot find helpful information at this point, they are likely to be unhappy and frustrated [9]. In addition, the size labels provide insufficient information to guide consumers in the initial selection of garments [10]. Recently, Faust and Carrier [1] concluded their study by mentioning that standardization of size labels is needed in order to provide better fitting clothes for RTW.

In this century, when things are moving very fast and time is valuable, it is very time consuming for customers to try on clothing more than once. They are no longer willing to undertake multiple fittings. Researchers have recommended that to solve fit problems the size designation system should be based on anthropometric data [1,11–14]. Without proper size designations, consumers are more likely to purchase incorrectly sized garments and will have to repeatedly try on garments, which can cause damage to the merchandise, and the rate of returns will be higher [1].

With all the confusion of size labels and problems that apparel consumers face with RTW clothing, it has been suggested that manufacturers should follow one standard way of labeling clothing sizes [6,15]. In addition, one recommended solution is the Mondoform labeling system developed in 1975 [16]. Under the Mondoform system, sizes are clearly depicted by wordless pictograms that indicate the body dimensions that garments are designed to fit [17]. As suggested by many researchers, this resolves the problem of size labeling by defining the size of a garment in terms of the common control dimensions used by many standards: bust, waist, hips, and height [18,19].

9.2.1 Mondoform labeling

The sizing system is not complete until it is coded and labeled. Most ASTM standards label sizes with a number from 00 to 20. Some companies use bust measurements to label the sizes. Knitwear is often labeled as small, medium, and large. Top wear and bottom wear have different labeling systems. Outwear and underwear have different labeling systems. Consumers often get confused with all these labeling systems or methods. Faust and Carrier designed a labeling system with a graph illustrating the silhouette (Fig. 9.3) that would "not only convey better information but also be highly predictive of the garment (pants in this case) would that women find fitting difficult [20].

Figs. 9.2–9.5 are examples of Mondoform wordless pictograms using the standard EN 1342 in showing the key dimensions for size designation. Main examples of indications are shown in these figures as Figs. 9.2 and 9.3 depict the upper and lower body of the male sample population, and Figs. 9.4 and 9.5 depict the female sample population. The indicated values are the designation of sizes by range, which show the relevant garments can be worn by the wearer within the indicated size range.

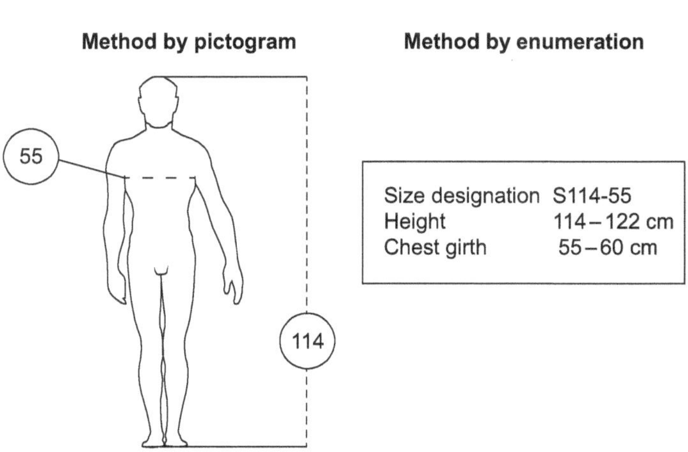

Figure 9.2 Mondoform labeling for male small size, upper body.

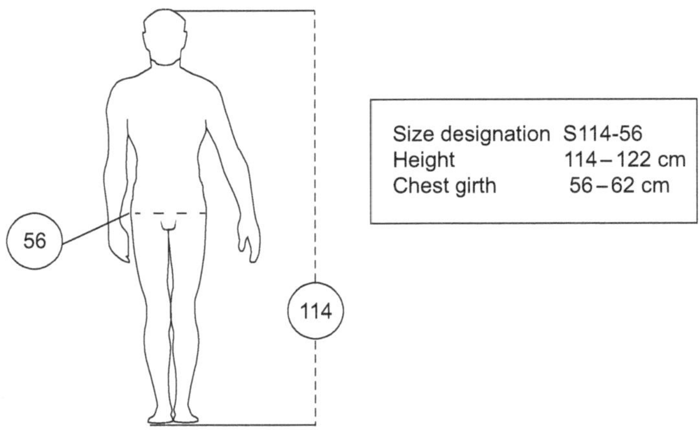

Figure 9.3 Mondoform labeling for male small size, lower body.

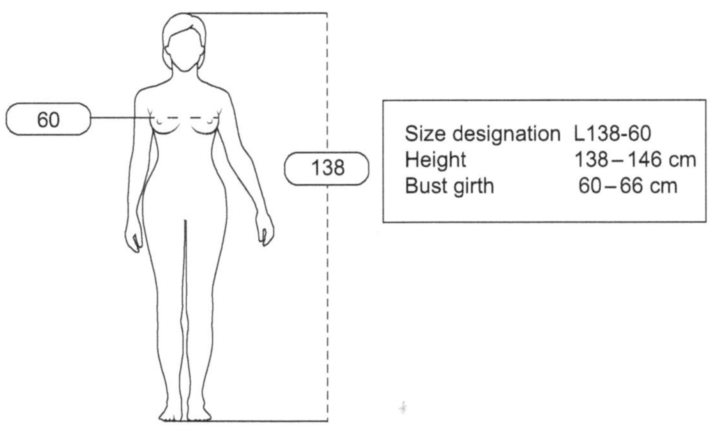

Figure 9.4 Mondoform labeling for female large size, upper body.

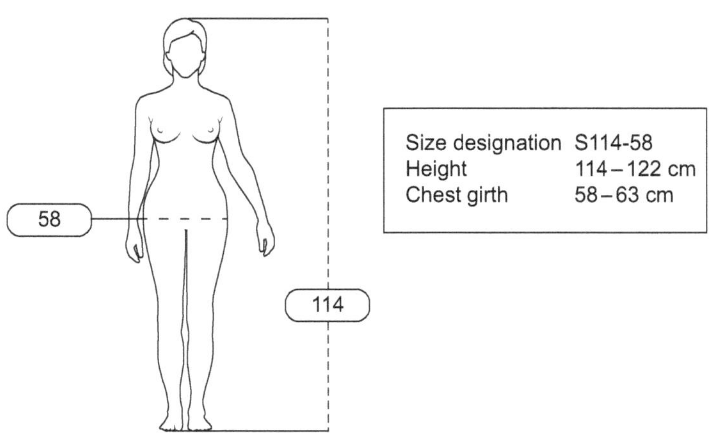

Figure 9.5 Mondoform labeling for female small size, lower body.

9.3 Validation of sizing designation and number of sizes

Based on the analysis of Figs. 8.1–8.9, it confirms a considerable number of samples that meet the specifications of classified rules (see Tables 7.15 and 7.16) obtained from the decision tree technique for all the sample population according to different genders and age groups.

The evidence from this section also suggests that the size table has successfully accommodated the variations of body size and shape found in the sample size population as shown in Figs. 8.1–8.9. One apparent example can be seen in Fig. 8.9 where there are four options of height to choose from for females with 85 cm hip. If they are either 142 or 148 cm, then they are classified as small size. Furthermore, if they are either 154 or 160 cm with 85 cm hip, they are classified as medium sizes. Consequently, for females either in small or medium size with 85 cm hip, the sizes are different from each one. Practically, this means that there are four different size designations to accommodate females of 85 cm hip. This literally implies that this size table reveals the actual distribution of measurements obtained from the sample population.

In comparison to girth variation, the same Fig. 8.9 shows that there are variations of girth measurements for one height size. For example, a person who is 148 cm tall can be classified either in small or large sizes, which are two extreme sizes determined by different girth measurements. Furthermore, from Fig. 8.9, it can be seen that in small size, those with a height of 148 cm can have four different hip measurements, which implies the four different sizes.

In contrast, for the same height, samples can be classified under large size with two different hip sizes. In total, samples of those who are 148 cm tall have a selection of six different hip sizes to fit their body size and shapes. From these results, the size tables show the capability to accommodate the different sizes and shapes, and therefore this assists in understanding the role of anthropometric data in the development of an efficient sizing system. This also truly affirms that with the right size developed for the sample size, the right fit of garments can be achieved among the sample population. This study produced results that corroborate the significant findings of the previous work in this field [11,21].

The summary of the results are depicted in Table 9.1 and Table 9.2 for upper and lower body. These tables summarize the results of the size system according to the key dimensions in which the profile of size tables functions as the guideline to find the correct size. Both tables indicate the whole size range developed for the entire sample size in this study.

As can be seen from Table 7.15 and Table 7.16, the rules clearly depict the different sizes of the sample population for different genders (males and females) and age group (7–12 and 13–17). This confirms the well-defined cluster groups illustrated by the scatter plot in cluster analysis section. Moreover, it also reflects the results of the decision tree that demonstrated the high percentage correctness of the classified models.

Table 9.1 **Size table for upper body according to gender and age groups**

Size table for children (7–12 and 13–17): Upper body				
	Key dimensions	**Small size (*S*) (cm)**	**Medium size (*M*) (cm)**	**Large size (*L*) (cm)**
Male (7–12)	Height	114–129.9	130–153.9	130–153.9
	Chest girth	55–84.9	55–74.9	75–89.9
Female (7–12)	Height	114–129.9	114–129.9	138–153.9
	Bust girth	54–65.9	66–83.9	60–89.9
			130–137.9	
			54–83.9	
Male (13–17)	Height	146–157.9	158–175.9	146–175.9
	Chest girth	70–87.9	70–87.9	88–105.9
Female (13–17)	Height	142–153.9	154–165.9	142–165.9
	Bust girth	70–87.9	70–87.9	88–99.9

Table 9.2 **Size table for lower body according to gender and age groups**

Size table for children (7–12 and 13–17): Lower body				
	Key dimensions	**Small size (*S*) (cm)**	**Medium size (*M*) (cm)**	**Large size (*L*) (cm)**
Male (7–12)	Height	114–129.9	130–153.9	130–153.9
	Hip girth	56–85.9	56–73.9	74–91.9
Female (7–12)	Height	114–137.9	122–137.9	138–153.9
	Hip girth	58–67.9	68–97.9	63–97.9
		114–121.9		
		68–73.9		
Male (13–17)	Height	146–157.9	158–175.9	146–175.9
	Hip girth	71–90.9	71–90.9	91–105.9
Female (13–17)	Height	142–153.9	154–165.9	142–165.9
	Hip girth	75–94.9	75–94.9	95–104.9

For example, for upper body in male age 7–12, the sizes are revealed in three different sizes, namely small, medium, and large. It is obvious from the table that samples are clearly separated according to height and girth variables. The respective body sizes do not have common characteristics, and as such, it is easier for the customers to distinguish sizes and for manufacturers to select sizes to be manufactured.

However, although all sizes show clear differences between sizes, the medium size for female ages 7–12 for upper body shows somewhat complicated rules. As can be seen in Table 9.1, the medium size shows two rules instead of one typical rule to denote the size. It means that if the sample female is between 114 and 129.9 cm tall and has a bust measurement between 66 and 83.9 cm bust girth, then she is classified under medium size. In addition, if height is between 130 and 137.9 cm, and bust girth is between 54 and 89.9 cm, then she is also considered a medium size.

9.3.1 Male size designation tables (ages 7–12)

Tables 9.3–9.5 indicate the three body types—small, medium, and large—using the size designation tables for the upper body. Tables 9.6–9.8 are the three size designation tables that were produced for the lower body. All sizes are in cm.

9.3.2 Female size designation tables (ages 7–12)

Three size designation tables, Tables 9.9–9.11, are depicted for all the samples according to small, medium, and large body types for the upper body, and another three size designation tables, Tables 9.12–9.14 are depicted for the lower body. All sizes are in cm.

9.3.3 Male size designation tables (ages 13–17)

Three size designation tables, Tables 9.15–9.17 are depicted for all the samples in this group according to small, medium, and large body types for the upper body, and another three size designation tables, Tables 9.18–9.20 are depicted for the lower body. All sizes are in cm.

9.4 Female size designation tables (ages 13–17)

Tables 9.21–9.23 illustrate the upper body of females ages 13–17 according to small, medium, and large body types. Another three size designation tables, Tables 9.24–9.26 illustrate the lower body.

9.5 Children sizing designation potential application to different groups of children and teenagers

This research has demonstrated significant findings as the sample population was segmented according to height and girth variables. Body sizes and shapes were reflected in each body type obtained from cluster analysis technique. The result is found differently if using height as the primary control dimension. This can be seen in Figs. 9.6 and 9.7. These size tables that denote the sizing designation were developed based on the same sample group but using different primary control dimensions. In Fig. 9.6, the size system was developed using height as the primary control dimension. In contrast, Fig. 9.7 uses the height and chest girth as the primary control dimensions.

Table 9.3 Small size for male (7–12 years old) upper body

Designation		S114-55	S114-60	S114-65	S114-70	S122-55	S122-60	S122-65	S122-70	S122-75	S122-80
Control dimensions	Height	114–122	114–122	114–122	114–122	122–130	122–130	122–130	122–130	122–130	122–130
	Chest	55–60	60–65	65–70	70–75	55–60	60–65	65–70	70–75	75–80	80–85

Table 9.4 Medium size for male (7–12 years old) upper body

Designation		M130-55	M130-60	M130-65	M130-70	M138-60	M138-65	M138-70	M146-65	M146-70
Control dimensions	Height	130–138	130–138	130–138	138–146	138–146	138–146	138–146	146–154	146–154
	Chest	55–60	60–65	65–70	70–75	60–65	65–70	70–75	65–70	70–75

Table 9.5 Large size for male (7–12 years old) upper body

Designation		L130-75	L130-80	L130-85	L138-75	L138-80	L138-85	L146-75	L146-80	L146-85
Control dimensions	Height	130–138	130–138	130–138	138–146	138–146	138–146	146–154	146–154	146–154
	Chest	75–80	80–85	85–90	75–80	80–85	85–90	75–80	80–85	85–90

Table 9.6 Small size for male (7–12 years old) lower body

Designation		S114-56	S114-62	S114-68	S114-74	S122-56	S122-62	S122-68	S122-74	S122-80
Control dimensions	Height	114–122	114–122	114–122	114–122	122–130	122–130	122–130	122–130	122–130
	Chest	56–62	62–68	68–74	74–80	56–62	62–68	68–74	74–80	80–86

Table 9.7 Medium size for male (7–12 years old) lower body

Designation		M130-56	M130-62	M130-68	M138-62	M138-68	M146-68
Control dimensions	Height	130–138	130–138	130–138	138–146	138–146	146–154
	Chest	56–62	62–68	68–74	62–68	68–74	68–74

Table 9.8 Large size for male (7–12 years old) lower body

Designation		L130-74	L130-80	L130-86	L138-74	L138-80	L138-86	L146-74	L146-80	L146-86
Control dimensions	Height	130–138	130–138	130–138	138–146	138–146	138–146	146–154	146–154	146–154
	Chest	74–80	80–86	86–92	74–80	80–86	86–92	74–80	80–86	86–92

Table 9.9 **Small size for female (7–12 years old) upper body**

Designation		S114-54	S114-60	S122-54	S122-60
Control dimensions	Height	114–122	114–122	122–130	122–130
	Bust	54–60	60–66	54–60	60–66

From Fig. 9.6, when samples were segregated according to height, the sizes created are reflected on human body sizes (length) where *S* shows short people, *M* reflects average height people, and *L* reflects tall people. Thus, each cluster group was based on height category with different chest variables. As compared to Fig. 9.7, dividing samples according to height and chest, apparent characteristics of shapes and sizes can be seen in each cluster group where green *S* shows petite groups, *M* reflects slender people, and *L* reflects large people. This proof segregating the samples using both girth and length together will reveal the different body sizes and shapes of the samples.

The significance of reflecting the body sizes and shapes in a sizing system is to the benefit of manufacturers. Consequently, manufacturers are able to design clothing according to different body types because different sizes and shapes need different kinds of design to enhance the look of the person. For example, in the large size group, design of clothing needs to conceal the shape as compared to slender people where design can enhance the beauty. Moreover, it has been proven by previous studies that size and design of clothing are equally significant in the selection of clothing. Therefore, by developing sizing systems that differentiate the body sizes and shapes will enhance the customers' appearances and benefit the manufacturers and retailers.

Decision tree technique effectively classified the sample groups according to clusters, which established the profiles for each sample group. The eight models attained for the sample groups describing the upper and lower body yielded better than 85% correctness accuracy. The most significant finding is the profile model, which gives 24 classification rules for classifying the samples according to body type.

Size tables developed for each sample group using the established classified rules effectively accommodated the variations of body size and shape found in the sample size population. Twenty-four size tables describing the upper and lower body of each gender and age group were successfully developed. The 24 size tables were aggregated into two summary tables (Tables 6.2 and 6.3) to describe the complete size roll for the entire sample in this study. These tables are useful as a quick guideline in order to determine to which size a given individual belongs.

To distinguish the efficiency and accuracy of the proposed sizes, the sizes are then validated using the aggregate loss of fit, size rolls, and coverage rate criteria. The findings show that the aggregate loss for all sizes was good (within 2.5–3.2 cm), and therefore the differences between the assigned sizes and the actual sizes are lower than the threshold required. This indicates that the proposed sizes established from the sample population are accurate. In addition, this study yielded fewer sizes compared

Table 9.10 **Medium size for female (7–12 years old) upper body**

Designation		M114-66	M122-66	M122-72	M122-78	M130-54	M130-60	M130-66	M130-72	M130-78	M130-84
Control dimensions	Height	114–122	122–130	122–130	122–130	130–138	130–138	130–138	130–138	130–138	130–138
	Bust	66–72	66–72	72–78	78–84	54–60	60–66	66–72	72–78	78–84	84–90

Table 9.11 **Large size for female (7–12 years old) upper body**

Designation		L138-60	L138-66	L138-72	L138-78	L138-84	L146-60	L146-66	L146-72	L146-78	L146-84
Control dimensions	Height	138–146	138–146	138–146	138–146	138–146	146–154	146–154	146–154	146–154	146–154
	Bust	60–66	66–72	72–78	78–84	84–90	60–66	66–72	72–78	78–84	84–90

Table 9.12 **Small size for female (7–12 years old) lower body**

Designation		S114-58	S114-63	S114-68	S122-58	S122-63	S130-58	S130-63
Control dimensions	Height	114–122	114–122	114–122	122–130	122–130	130–138	130–138
	Hip	58–63	63–68	68–74	58–63	63–68	63–68	58–63

Table 9.13 **Medium size for female (7–12 years old) lower body**

Designation		M122-68	M122-74	M122-80	M122-86	M130-68	M130-74	M130-80	M130-86	M130-92
Control dimensions	Height	122–130	122–130	122–130	122–130	130–138	130–138	130–138	130–138	130–138
	Hip	68–74	74–80	80–86	86–92	68–74	74—80	80–86	86–92	92–98

Table 9.14 **Large size for female (7–12 years old) lower body**

Designation		L138-63	L138-68	L138-74	L138-80	L138-86	L138-92	L146-68	L146-74	L146-80	L146-86	L146-92
Control dimensions	Height	138–146	138–146	138–146	138–146	138–146	146–154	146–154	146–154	146–154	146–154	146–154
	Hip	63–68	68–74	74–80	80–86	86–92	74–80	68–74	74–80	80–86	86–92	92–98

Table 9.15 **Small size for male (13–17 years old) upper body**

Designation		S146-70	S146-76	S146-82	S152-70	S152-76	S152-82
Control dimensions	Height	146–152	146–152	146–152	152–158	152–158	152–158
	Chest	70–76	76–82	82–88	70–76	76–82	82–88

Table 9.16 **Medium size for male (13–17 years old) upper body**

Designation		M158-70	M158-76	M158-82	M164-70	M164-76	M164-82	M170-76	M170-82
Control dimensions	Height	158–164	158–164	158–164	164–170	164–170	164–170	170–176	170–176
	Chest	70–76	76–82	82–88	70–76	76–82	82–88	76–82	82–88

Table 9.17 **Large size for male (13–17 years old) upper body**

Designation		L146-88	L146-94	L152-88	L152-94	L158-88	L158-94	L164-88	L164-94	L164-100	L170-88	L170-94	L170-100
Control dimensions	Height	146–152	146–152	152–158	152–158	158–164	158–164	164–170	164–170	164–170	170–176	170–176	170–176
	Chest	88–94	94–100	88–94	94–100	88–94	94–100	88–94	94–100	100–106	88–94	94–100	100–106

Table 9.18 **Small size for male (13–17 years old) lower body**

Designation		S146-71	S146-76	S146-81	S146-86	S152-71	S152-76	S152-81	S152-86
Control dimensions	Height	146–152	146–152	146–152	146–152	152–158	152–158	152–158	152–158
	Hip	71–76	76–81	81–86	86–91	71–76	76–81	81–86	86–91

Table 9.19 **Medium size for male (13–17 years old) lower body**

Designation		M158-71	M158-76	M158-81	M158-86	M164-76	M164-81	M164-86	M170-76	M170-81	M170-86
Control dimensions	Height	158–164	158–164	158–164	158–164	164–170	164–170	164–170	170–176	170–176	170–176
	Hip	71–76	76–81	81–86	86–91	76–81	81–86	86–91	76–81	81–86	86–91

Table 9.20 **Large size for male (13–17 years old) lower body**

Designation		L146-91	L152-91	L152-96	L158- 91	L158-96	L164-91	L164-96	L164-101	L170-91	L170-96	L170-101
Control dimensions	Height	146–152	152–158	152–158	158–164	158–164	164–170	164–170	164–170	170–176	170–176	170–176
	Hip	91–96	91–96	96–101	91–96	96–101	91–96	96–101	101–106	91–96	96–101	101–106

Table 9.21 **Small size for female (13–17 years old) upper body**

Designation		S142-70	S142-76	S142-82	S148-70	S148-76	S148-82
Control dimensions	Height	142–148	142–148	142–148	148–154	148–154	148–154
	Bust	70–76	76–82	82–88	70–76	76–82	82–88

Table 9.22 **Medium size for female (13–17 years old) upper body**

Designation		M154-70	M154-76	M154-82	M160-70	M160-76	M160-82
Control dimensions	Height	154–160	154–160	154–160	160–166	160–166	160–166
	Bust	70–76	76–82	82–88	70–76	76–82	82–88

Table 9.23 **Large size for female (13–17 years old) upper body**

Designation		L142-88	L148-88	L148-94	L154-88	L154-94	L160-88	L160-94
Control dimensions	Height Bust	142–148 88–94	148–154 88–94	148–154 94–98	154–160 88–94	154–160 94–98	160—166 88–94	160–166 94–98

Table 9.24 **Small size for female (13–17 years old) lower body**

Designation		S142-75	S142-80	S142-85	S142-90	S148-75	S148-80	S148-85	S148-90
Control dimensions	Height Bust	142–148 75–80	142–148 80–85	142–148 85–90	142–148 90–95	148–154 75–80	148–154 80–85	148–154 85–90	148–154 90–95

Table 9.25 **Medium size for female (13–17 years old) lower body**

Designation		M154-75	M154-80	M154-85	M154-90	M160-80	M160-85	M160-90
Control dimensions	Height Bust	154–160 75–80	154–160 80–85	154–160 85–90	154–160 90–95	160–166 80–85	160–166 85–90	160–166 90–95

Table 9.26 **Large size for female (13–17 years old) lower body**

Designation		L142-95	L148-95	L148-100	L154-95	L154-100	L160-95	L160-100
Control dimensions	Height Bust	142–148 95–100	148–154 95–100	148–154 100–105	154–160 95–100	154–160 100–105	160–166 95–100	160–166 100–105

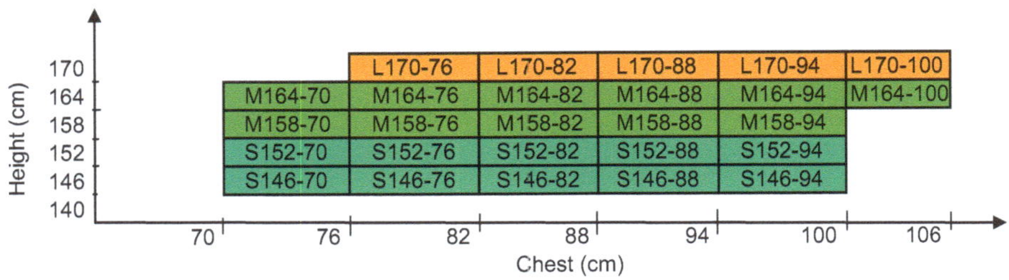

Figure 9.6 Sizes obtained when clustered according to height as the primary control dimension.

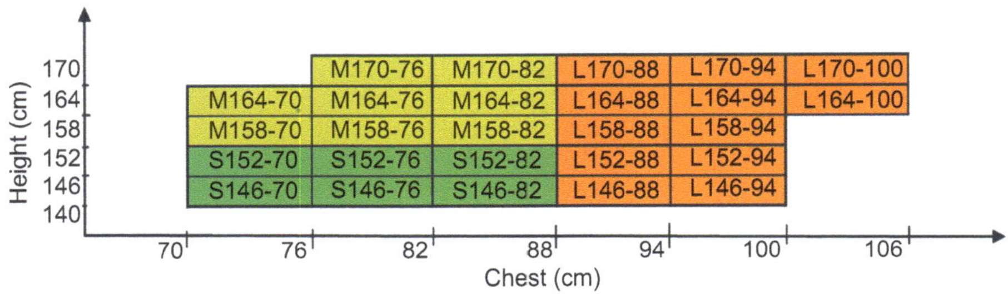

Figure 9.7 Sizes obtained when clustered according to height and chest as the primary control dimensions.

to previous studies. Not only that, the sizes also accommodated an average of 98% of the sample population, which indicates high efficiency. Therefore, the sizing systems established for each sample group based on upper and lower body are proven effective based on the validation criteria.

Based on the developed sizes, meaningful size designation tables following the standard EN 13402-3 were established for the entire sample population, and to complement the sizing system a proper size label that carries useful size information was created using the Mondoform standard labeling technique. Both the tables and label reflect the key dimensions; this provides easier identification of sizes and therefore may assist in the selection of the right sizes for a better fit.

Overall, the proposed classification sizes allowed almost all the samples (average of 98%) into the size designation, which proves the accuracy and high efficiency of the newly created sizing system based on the anthropometric data of the targeted population.

9.6 Conclusions

This research has successfully developed sizing systems for male and female children ages 7–17 years old based on anthropometric data. These children were clustered into different age groups and genders for four groups, namely females ages 7–12, males ages 7–12, females ages 13–17, and males ages 13–17. Consequently, eight sizing systems were developed to define upper and lower body sizes for each of the four sample populations.

A standard sizing system is essential for addressing the range of sizes that exist within the population of children in a particular retail market. Recognizing the need for a standard sizing system based on anthropometric data, a comprehensive anthropometric survey was successfully conducted measuring 2500 children in order to collect the data. Analysis of the anthropometric data collected revealed that anthropometric differences between genders do exist, and significant differences between anthropometric data for different ages were revealed. Therefore, this study's findings substantiate the need to develop a sizing system according to different gender and age groups.

The body types found in the sample population can be summarized as follows:

1. small-sized body type with short height and small girths or short height with small-to-large girths
2. medium-sized body type with average height and small girths or average height with average-to-large girths
3. large-sized body types with tall height and small girths, tall height with medium-to-large girths, and tall with small-to-large girths

9.7 Recommendations for future work

The present study proposed new clothing sizes for a sample population of children ranging from 7 to 17 years old. In order to further enhance the study of fit in clothing, the following suggestions are provided for future research.

The focus of this study was on children in only one state in a country. Findings could be more generalized by getting data from a more representative sample of the population. Specifically, a survey should cover all regions of the country such that the resulting sizing system would represent a larger sample of the population of children.

It is also important to consider different market segments for a new standard sizing system. For instance, in order to enhance the sizing system, the same research could be replicated using different populations of children such as toddlers, infants, plus sizes, and special-needs children.

This study used the manual method of measuring the body for the collection of anthropometric data. However, with the advancement of technology, many studies in sizing research have moved toward collecting the data using 3D body scanners. Future research could adopt the same technique using 3D body scanning technology to facilitate data collection and reduce the time involved.

Furthermore, the usage of different analysis techniques such as artificial neural network and other advanced computer programming techniques could be explored in order to produce the sizing system that can give different size results and efficiency.

References

[1] M.E. Faust, S. Carrier, A proposal for a new size label to assist consumers in finding well-fitting women's clothing, especially pants: an analysis of size USA female data and women's ready-to-wear pants for North American companies, Text. Res. J. 79 (16) (2009) 1446–1458.
[2] S.L. Amirul Ihsan, Beauty: one size doesn't fit all, in: New Strait Times, April 27th. Kuala Lumpur, 2009.
[3] J.C. Yoon, C.R. Jasper, Garment-sizing systems: an international comparison, Int. J. Cloth. Sci. Technol. 5 (1993) 28–37.
[4] J.C. Yoon, C. Jasper, Key dimensions of women's ready-to-wear apparel: developing a consumer size-labeling system, Cloth. Text. Res. J. 14 (1996) 89–95.
[5] M.E. Faust, S. Carrier, P. Baptiste, Variations in Canadian women's ready-to-wear standard sizes, J. Fash. Market. Manag. 10 (2006) 71–83.
[6] C.J. Yoon, C.R. Jasper, Consumer preferences for size description systems of men's and women's apparel, J. Consum. Aff. 29 (1995) 429–442.
[7] L.D. Burns, N.O. Bryant, The Business and Fashion Designing, Manufacturing and Marketing, 2002.
[8] J.E. Workman, S.E. Lentz, Measurement specifications for manufacturers' prototype bodies, Cloth. Text. Res. J. 18 (2000) 251–259.
[9] L. Kirsi, B. Hauge, I.G. Klepp, Large? Clothing Size and Size Labeling, TemaNord 2009:503, Nordic Council of Ministers, Copenhagen, 2009.
[10] D. Ujevic, R. Hrzenjaka, K. Dolezal, et al., Size designation system of clothes and footwear based on Croatian anthropometric system [J], J. Fiber Bioeng. Inf. 4 (4) (2011) 311–319.
[11] A. Petrova, Creating sizing systems, in: S.P. Ashdown (Ed.), Sizing in Clothing Developing Effective Sizing Systems for Ready-to-Wear Clothing, Woodhead Publishing, Cambridge, 2007, pp. 57–82.
[12] A.M. Mason, D. Klerk, S.P. Ashdown, Sizing and fit research at grassroots level. A methodology for the identification of unique body shapes in African developing countries, J. Fam. Ecol. Consume. Sci. 36 (2008) 9–21.
[13] D. Gupta, Anthropometric data analysis and garment sizing system for Indian population, in: Presented at 86th Textile Institute Conference, Hong Kong, 2008.
[14] R. Otieno, C. Fairhurst, The development of new clothing size charts for female Kenyan children. Part I: using anthropometric data to create size charts, J. Text. Inst. 1754–2340 91 (2000) 143–152.
[15] G.N. Darko Ujevi, K. Doležal, L. Szirovicza, New anthropometric instruments, Coll. Antropol. 31 (2007) 1031–1038.
[16] T.R. Kinley, Size variation in women's pants, Cloth. Text. Res. J. 21 (2003) 19–31.
[17] ISO standard, International Standards Organization, ISO/TR10652, Standard Sizing Systems for Clothes, 1991.
[18] K.L. LaBat, Consumer Satisfaction/Dissatisfaction with the Fit of Ready-to-Wear Clothing (Ph.D. dissertation), University of Minnesota, 1987.
[19] J.M. Winks, Clothing Sizes: International Standardization, 1997.
[20] S. Carrier, M.E. Faust, Women's wear sizing: a new labeling system, J. Fash. Market. Manag. 14 (1) (2010) 88–126.
[21] D. Gupta, N. Garg, K. Arora, N. Priyadarshini, Developing body measurement charts for garment manufacture based on a linear programming approach, J. Text. Appar. Technol. Manag. 5 (2006) 1–13.

Index

'*Note:* Page numbers followed by "f" indicate figures, "t" indicate tables.'